37th European Peptide Symposium 2024 (EPS 37)

Florence, Italy
25-29 August 2024

ISBN: 979-8-3313-1304-3

Printed from e-media with permission by:

Curran Associates, Inc.
57 Morehouse Lane
Red Hook, NY 12571

Some format issues inherent in the e-media version may also appear in this print version.

Printed with permission by Curran Associates, Inc. (2025)

For permission requests, please contact European Peptide Society
at the address below.

European Peptide Society
c/o Professor Paula Gomes
Rua do Campo Alegre, s/n
PT-4169-007 Porto
Portugal

Phone: +351 220 402 563

pgomes@fc.up.pt

Additional copies of this publication are available from:

Curran Associates, Inc.
57 Morehouse Lane
Red Hook, NY 12571 USA
Phone: 845-758-0400
Fax: 845-758-2633
Email: curran@proceedings.com
Web: www.proceedings.com

37th EPS
European Peptide Symposium

14th IPS
International Peptide Symposium

25 - 29 August 2024

PROCEEDINGS

Florence, Italy

http://doi.org/10.17952/37EPS

List of Lectures

List of Posters ****MAJORITY OF POSTERS UNAVAILABLE****

P1002 Convergent synthesis of peptidomimetic based on a chloroalkene dipeptide isostere and its application to inhibition against amyloid β aggregation
Takuya Kobayakawa, Chika Azuma, Kenichi Yamamoto, Kohei Tsuji, Hirokazu Tamamura
P1003 Optimization of IEDDA reaction parameters
Rasheda Aktar Samiha, Samir Das, Arundhati Nag
P1004 Identification of reactivity hotspots at the interfaces of protein–protein interactions
Nathalie M. Grob, Clint Remarcik, Simon L. Rössler, Jeffrey Y. K. Wong, John C. K. Wang, Jason Tao, Corey L. Smith, Andrei Loas, Stephen Buchwald, Dan Eaton, Magdalena Preciado López, Bradley Pentelute
P1005 Using N-acylation as a disulfide lock simplifies chemical protein synthesis
Vincent Diemer, Vangelis Agouridas, Benoît Snella, Rémi Desmet, Oleg Melnyk
P1007 Synthesis of Peptide Hydrazides: Hydrazone Resin vs. Hydrazine 2CT Resin
Maria Leko, Polina Filippova, Karin Rustler, Thomas Bruckdorfer, Sergey Burov
P1009 Facilitating the access to repeat proteins by a novel water-compatible oligomerization process
Eliott Roy, Oleg Melnyk, Vincent Diemer
P1012 Technetium-99m tricarbonyl complexes containing amino acid derivatives for cancer imaging
João D.G. Correia, Rúben D.M. Silva, Cristina P. Matos, Catarina I.G. Pinto, Joana F. Guerreiro, Lurdes Gano, Filipa Mendes
P1013 Development of methodology for late-stage acylation of threonine residue enabled by chemoselective acyl transfer followed by desulfurization
Daiki Sato, Chiho Shinozaki, Masaya Denda, Akira Otaka
P1014 Towards the total synthesis of the bacteriocin Nisin A on solid support
Ilia Perov, Bethan Donnelly, Tyler Mallett, John Vederas
P1015 Protein semi-synthetic approach to probe C-terminal hyperphosphorylation of the Alzheimer-relevant protein Tau
Bingjia Yan, Léa El Hajjar, Wenyi Lia, Caroline Smet-Nocca, Christian P. R. Hackenberger
P1016 Structural Optimization Study Leading to Orally Bioavailable Cyclic Peptides Inhibiting an Intracellular Tough Target
Mikimasa Tanada, Ryuji Hayashi, Mirai Kage, Atsushi Ohta, Hitoshi Iikura,Takuya Shiraishi
P1017 Chemical Engineering of Artificial Transcription Factors
Omer Harel, Xiaoxi Lin, Muhammad Jbara
P1018 Peptides Targeting DNA Four-Way Junctions
Eleanor Ivens, Andrew Beekman, Mark Searcey
P1019 C-Glycosylation of Aromatic Amino Acids by Pd-Catalyzed Cross-Coupling
Calvin Matula, Norbert Sewald
P1020 Efficient Peptide Thioester Synthesis from Hydrazides Using Near-Stoichiometric Thiol Quantities
Julie Di Adamo, Nathalie Ollivier, Vincent Diemer, Oleg Melnyk
P1021 Electrostatic Assistance of NCL Catalysis by 4-Mercaptophenylacetic Acid
Nathalie Ollivier, Eliott Roy, Remi Desmet, Vangelis Agouridas, Vincent Diemer, Oleg Melnyk
P1022 An efficient strategy to bioconjugate disulfide-rich peptides via thiol-maleimide chemistry while preserving their bridging pattern
Lylia Azzoug, Hervé Meudal, Karine Loth, Séverine Morisset-Lopez, Carlo Pifferi,Vincent Aucagne
P1023 Auxiliaries in Native Chemical Ligation: Exploring the Reactivity Landscape with new Scaffolds
Sebastian Trunschke, Antonia Feida, Oliver Seitz
P1025 Superfast Desulfurization for Protein Chemical Synthesis and Modification
Zhenquan Sun, Yihui Cao, Xuechen Li
P1026 An iron-catalyzed protein desulfurization method reminiscent of aquatic chemistry
Rémi Desmet, Céline Wichlacz, Aurélie Tasiemski, Vangelis Agouridas, Oleg Melnyk
P1027 Chemoselective P-S coupling of aminophospholes to peptides under mild conditions
Emmanuelle Rémond, Laure Liénart, Miyanou Rosales-Hurtado, Sébastien Clément, Jean-Alain Fehrentz, Jean-Louis Banères, Florine Cavelier
P1029 Flow – SPPS towards greener and more efficient peptide manufacturing
Oliver Kracker, Eike-Fabian Sachs
P1030 Design of Experiments as Powerful Tool for Peptide Cleavage Development
Teresa Friedrichs, Alexander P. Galow

P1031 Synthetic combinatorial OPA-cyclic peptide library
Kin Nam Adrian Chu, Carina Hey Pui Cheung, Xuechen Li
P1032 Direct synthesis of peptidomimetics containing non-coded amino acids via photocatalytic carbamoylation reaction
T. Gandini, M. L. Gelmi, R. Bucci
P1034 Thiol-yne Mediated Macrocyclisation and Derivatisation Strategies
Conor Williams, Dearbhla Tully, Susannah H. Calvert, Joanna F. McGouran, Eoin M. Scanlan
P1036 Synthesis and biological activity of ultra-short antimicrobial peptides bearing a novel noncoded amino acid
Barbara Biondi, Fernando Formaggio, Cristina Peggion, Andrea Schivo, Teodora Calin, Simona Oancea, Emma Khachatryan, Liana Hayriyan, Ashot Saghyan, Anna Mkrtchyan
P1037 Evaluation of the stability of oxytocin and vasopressin under oxidative stress conditions
Stefano Tartaggia, Claudia Honisch, Martina Rotondo, Paolo Ruzza
P1038 Biocatalytic Synthesis of Dihalogenated Tryptophan Derivatives
Bjarne Scharkowski, Nicolai Montua, Norbert Sewald
P1039 Silicon-containing amino acids for 18F labeling of neurotensin analogues targeting NTS1 receptor
Miyanou Rosales Hurtado, Emmanuelle Rémond, Delphine Vimont, Alexis Reinbold, Elif Hindié, Clément Morgat, Florine Cavelier
P1042 Preclinical Development of Myrcludex B Derivatives: Physicochemical Properties
Tyra Christian, Sabrina Wohlfart, Christian Kleist, Philipp Uhl, Stephan Urban, Walter Mier
P1044 Exploring the structure-function relationship of heme binding to blood coagulation factor VIII based on peptide models
Lorena Kröner, Diana Imhof, Marie-T. Hopp
P1046 Site-specific post-translational modifications of Tau4
Miriam M. K. Preissler, Dominik Vogl, Christian F. W. Becker
P1047 Synthesis of the natural peptide product Fusahexin
Carmen Bäuerlein, Armin Geyer
P1049 Organocatalytic Surfactants for Direct Asymmetric Aldol Reactions in Aqueous Media
Susanta Kumar Behera, Ramesh Ramapanicker
P1050 O-to-O Acyl Transfer for Epimerization-free Peptide C-terminal Salicylaldehyde Ester Synthesis
Zhixiang Zhong, Wang Xia, Bing-Wen Li, Han Liu, Zhi-Xiang Yu, Xuechen Li
P1051 New electrochemical method for Arylation of Cysteine containing peptides
Shen Linhua, Olivier Monasson, Elisa Peroni, Franck Le Bideau, Samir Messaoudi
P1052 N-Aminoimidazol-2-one Peptide Mimic Tools for Deciphering Active Conformation 58
Yousra Hamdane, Pradeep Chauhan, Suresh Vutla, Darince Truong, William D. Lubell
P1054 Ring Pucker Control in β-Prolines 60
Nassim Maarouf Mesli, William D. Lubell
P1055 Nature-Inspired Gallinamides Are Potent Antischistosomal Agents: Inhibition of the cathepsin B1 Protease Target and Binding Mode Analysis
P. Spiwokova, M. Horn, L. Maresova, J. Fanfrlik, A. Jilkova, P. Fajtova, A. Leontovyc, R. Houstecka, L. Bielikova, J. Brynda, M. Chanova, H. Mertlikova-Kaiserova, E. Caro-Diaz, J. Almaliti, N. El-Sakkary, W.H. Gerwick, C.R. Caffrey, M. Mares
P1056 Study on interactions of LL-37 human cathelicidin peptide fragment with Cu(II) ion
Jakub Brzeski, Dariusz Wyrzykowski, Bogusław Pilarski, Joanna Makowska
P1058 Peptide stapling - application in antibacterial therapies
Monika Wojciechowska, Julia Macyszyn, Michal Burmistrz, Joanna Miszkiewicz, Malgorzata Lobka, Joana Trylska
P1059 Translocation-Enhancing Peptides: FacilitateTransport of Molecules and Lipid Scrambling
Ladislav Bartoš, Robert Vácha
P1062 Antimicrobial activity of Histatin 5 based metallopeptides
Gabriela Vieira Silva Zolin, Renan Lira de Farias, Saulo Santesso Garrido
P1063 De novo design of miniproteins binding to PD-L1
Juan Lizandra Pérez, Natalia Miodowska, Łukasz Berlicki
P1064 Metal complexes of ATCUN-like peptides for sequence-specific DNA cleavage
Ivan N. Unksov, Lena K. Müller, Hanna Zdhannova, Alexander Gräwe, Joshua Stahl, Viktor Stein, Daniel Tietze, Alesia A. Tietze
P1067 Development of peptide-based inhibitors against β-catenin using in silico approach
Minami Fujita, Amika Kikuchi, Kyohei Arita, Takashi Misawa, Yosuke Demizu

P1099 Biological Evaluation of New Fell Analogues as Potential PDE4 Inhibitors
Veronica Nemska, Boryana Borisova, Marie Cochard, Stéphane Gérard, Dancho Danalev
P1100 Investigation of the role of Mfn2 and Drp1 protein-protein interactions in mitochondrial dynamics and its biological implications in cardiovascular diseases
Berhe Hayelom, Qvit Nir
P1101 Peptide and Probe: A two-pronged approach to investigating menaquinone binding peptides
Eilidh J. Matheson, Roy A. M. van Beekveld, Sarah Gothard, Paolo Innocenti, Nathaniel I. Martin, Markus Weingarth, Stephen A. Cochrane
P1104 A series of novel, selective matriptase inhibitors
Dębowski Dawid, Aleksandra Helbik -Maciejewska, Agata Gitlin-Domagalska, Malihe Hassanzadeh, Antoine Désilets, Pierre-Luc Boudreault, Richard Leduc, Anna Łęgowska, Krzysztof Rolka
P1105 Total synthesis and semi-synthesis of fluorinated analogues of the antifungal cyclic lipopeptide iturin A
Periklis Karamanis, Matthew Kiernan, Jimmy Muldoon, Paul Evans, Cormac D. Murphy, Marina Rubini
P1106 Design and antimicrobial activity of short and novel peptides
Ilaria Di Stefano, Silvia Rizzo, Dario Corbisiero, Francesca Bonvicini, Giovanna Angela Gentilomi, Walter Cabri, Tainah Dorina Marforio, Lucia Ferrazzano, Alessandra Tolomelli
P1109 Short Analogues Of Cathelicidins Pmap-36 And Bmap-27: Structure-Activity Relationship And Stability To Proteolysis
C. Peggion, A. Schivo, F. Albini, A. Di Stasi, B. Biondi, M. Mardirossian
P1110 HBP-5: A Leading Heparin-Binding Peptide with Remarkable Activity against Gram-negative Bacteria
Roberto Bello Madruga, Daniel Sandín, Javier Valle, David Andreu, Marc Torrent
P1112 Inhibitory activity and proteolytic stability in human serum of peptide inhibitors of the Vascular Endothelial Growth Factor A165/Neuropilin-1 complex
Dagmara Tymecka, Patrycja Redkiewicz, Piotr F.J. Lipinski, Aleksandra Misicka
P1113 Unlocking the Potential of Conotoxins: Chemical Synthesis Strategies and Biological Activity
Yazid Souf, Nicolas Inguimbert, Sebastien Dutertre
P1114 Mucroporin-M1 analogues: synthesis, conformation and structure-activity correlation
Claudia Storti, C. Peggion
P1115 Ferrocene-peptide conjugates with potential antitumor activity
Barbara Biondi, Beatrice Calderaro, Annalisa Bisello, Roberta Cardena, Renato Schiesari, Saverio Santi
P1116 Fragments of fibroblast growth factors useful in regenerative medicine components
Katarzyna Czerczak-Kwiatkowska, Adrianna Wierzbicka, Dorota Bociaga, Justyna Fraczyk, Beata Kolesinska
P1117 CR22DK-S18G Breakthrough: Revealing a novel antimicrobial peptide with rifampicin and isoniazid-like potency against Mycobacterium tuberculosis - a comprehensive exploration from synthesis to immune modulation and selective action
Cesar Augusto Roque-Borda, Laura Maria Duran Gleriani Primo, Rafaela Melo Pogianeli, Lucas Henrique Domingos da Silva, Isabela Caroline dos Santos, Flávia Aparecida Resende, Henrik Franzyk, Paul Robert Hansen, Fernando Rogério Pavan
P1119 In silico identification and chemical remodelling of tick protein epitopes for vaccine antigen development
Stepan S. Denisov, M. Amine Jmel, Wouter Peeters, Maya Idema, Michalis Kotsyfakis, Johannes H. Ippel, Tilman M. Hackeng, Ingrid Dijkgraaf
P1121 Harnessing GPCRs Biased Signaling: Novel Macrocyclic Analogs for Enhanced Cardiovascular Treatment with Reduced Side Effects
Kien Tran, Xavier Sainsily, Jérôme Côté, David Coquerel, Pierre Couvineau, Sabrina Saibi, Lounès Haroune, Élie Besserer-Offroy, Alexandre Murza, Jean-Michel Longpré, Mannix Auger-Messier, Michel Bouvier, Philippe Sarret, Pierre-Luc Boudreault
P1122 A Novel Regulation Method of mRNA by Recruiting Regnase-1 Using Binding Motif-Antisense Oligonucleotide Conjugates
Masayuki Fujii
P1124 Deoxythymidine-based antimicrobial peptidomimetics to combat drug-resistance
Eun Young Kim, Dinesh Kumar, Sung Tae Yang, Song Yub Shin
P1125 Fine-Tuning the Immune-Stimulatory and Cancer Cell Binding Properties of Immune System Engagers (ISErs)
Hannah Back, Manuel Felkl, Susanne Kremer, Claudia Altdorf, Ute Steinbusch, Christian FW Becker
P1126 Targeted Acidosis Mediated Delivery of Antigenic MHC-Binding Peptides
Joey J. Kelly, Emily T. Ankrom, Sarah E. Newkirk, Damien Thévenin, Marcos M. Pires

P1151 Analysis of Mannan (polymannose) Conjugated with 35-55 Immunodominant Epitope of Myelin Oligodendrocyte Glycoprotein for the Treatment of Multiple Sclerosis
Carmen Simal, Areti Gkika, Maria-Eleni Androutsou, Georgia Biniari, Alexios Aletras, Theodore Tselios
P1152 A Nanomolar Potency Macrocyclic Peptide Inhibitor of SARS-CoV-2 Spike through Deep Mutational Scanning
Antonius J.P. Hopstaken, Oliver J. Debski-Antoniak, Preeti Hooda, Daniel L. Hurdiss, Seino A.K. Jongkees
P1153 Intelligent wound dressing with antimicrobial properties
Sebastian Kersting, Sophia Ohnesorg, Markus von Nickisch-Rosenegk
P1154 High Throughput Phage Display Screening Identified Peptide Inhibitors Targeting the α-Synuclein for Parkinson's Disease Treatment
Haiyan Zhou, Zhaohui Liang, Chenyin Wang, Wai Lok Ngai, Michael K. Chan, Chaogu Zheng, Xuechen Li
P1155 Linear peptides as IgG-epitope mimetic for allergic immunotherapy against birch pollen
Alessandro Streuli, Lara Šošić, Marta Paolucci, Agathe Duda, Klaus Eyer, Pål Johansen, Christian Steuer
P1160 Discovery of a novel α-helical peptide targeting GTP-bound K-Ras(G12D)
Hannah Comfort, Sarah Ross, Jefferson Revell, Thomas Murray, Carole Urbach, Bethan Archer, Darerca Owen, Helen Mott
P1161 Peptide Inhibitors of Protein-Protein Interactions Targeted to Src Homology 2 Domains: A Combined Computational and Spectroscopic Design Approach
Paolo Calligari, Claudia Storti, Chiara Innamorati, Chiara Fulci, Alessia Damiani, Cristiano DI Stefano, Chiara Cocca, Cristina Peggion, Barbara Biondi, Gianfranco Bocchinfuso, Lorenzo Stella
P1162 Peptide-based vesicle isolation method for P. aeruginosa 80
Tania Henriquez, Ismael Castañón, Silvia Scali, Chiara Falciani
P1164 Antibacterial, Anti-inflammatory and Anti-biofilm Activities of Novel Short Antimicrobial Peptides Designed from the Cecropin A-Melittin Hybrid Antimicrobial Peptide, BP100
Eun Young Kim, Dinesh Kumar, Sang Hyun Park, Song Yub Shin
P1165 Study on the stability of new promising antimicrobial peptides against protease activities present in infection sites
Clelia Cortese, Giovanni Cappello, Laura Cresti, Virginia Niccolini, Marco Scocchi, Marialuisa Mangoni, Alessandro Pini
P1170 Revolutionizing Peptide Production: Introducing An Engineered TEV Protease
R. Guerrette, C. Schwarz, B. Wetzel
P1172 Spatio-temporal control of cellular differentiation for cartilage tissue engineering
Jules Simonin Garcia, Cécile Echalier, Emmanuel Belamie, Marie Morille, Gilles Subra
P1173 Development of a novel alternative method for liraglutide production
Gabriela R. Barredo-Vacchelli, Jésica A. Rodríguez, Joaquín A. Eloy, Silvia A. Camperi
P1174 Comparison of in vitro immunogenicity between δ-ctenitoxin-Pn2a and epitope-based immunogens
Jésica A. Rodríguez, Gabriela R. Barredo-Vacchelli, Ariadna M. Birocco, Agustín Blachman, Gerardo Acosta, Fernando Albericio, Graciela C. Calabrese, Silvia A. Camperi
P1176 Extracellular-Vesicle Catch-and-Release isolation System using a Net-charge Invertible Curvature-sensing peptide
Kenichi Kawano, Yuki Kuzuma, Koichi Yoshio, Kenta Hosokawa, Yuuto Oosugi, Takahiro Fujiwara, Fumiaki Yokoyama, Katsumi Matsuzaki
P1178 Multifunctional peptides for the synthesis of photothermally active gold nanoparticles
Alžbeta Runová, Helma Wennemers
P1179 Intracellular Peptide Library Screening for Covalent Transcription Factor Inhibitors
Andrew Brennan, Keith Vance, Scott Lovell, Jody Mason
P1182 Innovative microfluidic preparation of monodisperse peptidic nanoparticlesfor ultrasound guided gene delivery
Guy Gouarin, Vicentiu Moraru, Samir Cherkaoui
P1186 Environment Responsive Peptide Dimers Bind and Stabilize dsRNA
Niall M. McLoughlin, Marvin A. Albers, Estel Collado Camps, Jannik Paulus, Youri A. Ran, Saskia Neubacher, Sven Henning, Roland Brock, Tom N. Grossmann
P1187 Engineered oxalyl thioester-containing protein domains as a building block for synthetic biopolymers
Francesco Terzani, Simindokht Rostami, Chen Wang, Benoît Snella, Birgit Wiltschi, Oleg Melnyk, Vangelis Agouridas

P1225 Development of novel amphipathic stapled peptides as DDS carriers for intracellular delivery of nucleic acids
Motoharu Hirano, Hidetomo Yokoo, Makoto Oba, Takashi Misawa, Yosuke Demizu
P1227 Development of selective peptide-based radiopharmaceuticals for targeted therapy and diagnosis of high malignant cancers
Anna Giorgio, Nicola Salvarese, Annarita Del Gatto, Michele Saviano, Cristina Bolzati, Laura Zaccaro
P1228 Control of biological processes using metallopeptides
Jessica Rodriguez Villar
P1229 pH-Dependent Cellular Uptake of CPPs – Guanidinium versus OxyguanidiniumProline
Rahel Heeb, Laura Massaad, Helma Wennemers
P1230 Lipophilic Prodrug Charge Masking- insights into new method of peptide prodrugs synthesis conception
Agata Gitlin-Domagalska, Anna Olejnik, Natalia Ptaszyńska, Jaroslaw Ruczyński, Mladena Glavaš, Mikolaj Marszałek, Dawid Dębowski, Aleksandra Helbik-Maciejewska, Anna Łęgowska, Chaim Gilon, Krzysztof Rolka
P1231 Amphipathic Proline-Rich Cell Penetrating Peptides for Mitochondria Targeting
Adeline M. Schmitt, Helma Wennemers
P1233 Optimization of Anti-Vasoconstricting UT-II-Derived Biogels
Thibault Cauwenbergh, Juliana Dallagnol, David Chatenet, Charlotte Martin, Steven Ballet
P1234 Development of cell penetrating peptide-based ubiquitin ligase ligand for the development of PROTAC
Hidetomo Yokoo, Zhou Dongrui, Yosuke Demizu
P1236 Design and characterization of CD20 antigen surfaces for the selection of rituximab peptide mimics
Océane Ricloux, Chanantida Jongwohan, Jordan Cossu, Liliane Guerente, Didier Boturyn
P1240 Development of Novel Angiopep-2 Conjugates for Cancer Treatment

Gábor Mező, Levente Dókus, Diána Vári-Mező, Krisztina Kiss, Lilla Pethő, Ivan Ranđelović, József Tóvári, Szilvia Bősze
P1241 Systematic Determination of the Effects of Macrocyclization on Peptide Permeability into Mycobacteria
Rachita Dash, Irene Lepori, Han Zhang, Wonpil Im, M. Sloan Siegrist, Marcos M. Pires
P1242 Lipophilic Prodrug Charge masking strategy for novel oxytocin prodrugs production
Agata Gitlin-Domagalska, Natalia Ptaszyńska, Anna Olejnik, Jaroslaw Ruczyński, Mladena Glavaš, Dawid Dębowski, Aleksandra Helbik-Maciejewska, Anna Łęgowska, Chaim Gilon, Krzysztof Rolka
P1243 Drug-likeness of cyclic peptides beyond Rule of 5
Shojiro Shinohara, Mikimasa Tanada, Yuya Morita, Kazuhiko Nakano, Takuya Shiraishi, Hitoshi Iikura
P1244 POM-peptide hybrid against Alzheimer's Desease
Marta Giacomini, Claudia Honisch, Elisabetta Galluppo, Genny Orso, Haihong Yu, Mauro Carraro, Paolo Ruzza
P1245 The potential use of peptides in pancreatic cancer
Martina Rotondo, Giovanna Bossio, Sofia Parassia, Andrea Mattarei, Claudia Honisch, Ildikò Szabò, Lucia Biasutto, Paolo Ruzza
P1246 Evaluation of different spin labels for the study of protein-peptide interaction in cellular extract using Electron Paramagnetic Resonance spectroscopy techniques
Claudia Toso, Giovanni Loprete, Elisabetta Bergantino, Simone Fabbian, Massimo Bellanda, Sylvain Marque, Gerard Audran, Marina Gobbo, Marta De Zotti, Donatella Carbonera, Marco Bortolus, Marilena DI Valentin
P1247 Development of a cell penetrating peptide carrying a biological active HPV16 E6 inhibitory molecule for delivery into cervical cancer cells
Anna Lucia Tornesello, Luisa Dassi, Vincenza De Gregorio, Giorgia Imparato, Marilisa Leone, Lu Wuyuan, Noemy Starita, Andrea Cerasuolo, Luigi Buonaguro, FrancoMaria Buonaguro, Maria Lina Tornesello
P1249 The Antimicrobial Peptide SET-M33: In Vivo Efficacy and Toxicity in Different Animal Species and Encapsulation in Inhalable Polymeric Nanoparticles for Pulmonary Delivery
Laura Cresti, Giovanni Cappello, Clelia Cortese, Virginia Niccolini, Alessandro Pini
P1250 Affibody-mediated peptide nucleic acid pretargeting for delivery of cytotoxic payloads to HER2 positive carcinoma
Jacob Clinton, Kristina Westerlund, Amelie Eriksson Karlström
P1251 Structure-activity relationship study on the P4 structure of macrocyclic BACE1 inhibitors
Kazuya Kobayashi, Momoka Kawakita, Takuya Otani, Yasaunao Hattori, Shinya Oishi
P1253 Non-Enzymatic Protein Catalyses Amide Bond Formation
Nicolas Brauckhoff, Laura Fang, Anissa Haim, Tom N. Grossmann
P1254 β-Sheet mimetics that Target the Transcriptional Coactivator β-Catenin
M. Wendt, A. Yeste-Vázquez, T. van Ramshorst, R. J. Hana, N. M. Pearce, S. Hennig, T. N. Grossmann

P1314 Self-assembled Peptides for Biomimetic Mineralization and Enamel Repair
Yue Li, Rong Chang, Yang-Jia Liu, Feng Chen, Yong-Xiang Chen
P1315 Cationic amphiphilic peptide-based delivery platforms for negatively charged active pharmaceutical ingredients including short interfering RNA
M. Rosa, C. Diaferia, A. Accardo, G. Morelli
P1316 Synthesis of peptidomimetics using Diaryl-β2,3 amino acids for the preparation of smart materials
Dulce A. Quintana-Romero, Raffaella Bucci, Merve Gul, Maria M. Pérez-Madrigal, Carlos Alemán, M. Luisa Gelmi
P1317 Conformational and self-assembly studies of a protein-mimetic peptide from Pseudomonas aeruginosa's YeaZ protein
Elettra Fasola, Giulia Alboreggia, Stefano Pieraccini, Francesco Oliva, Fatima Ezzahra Agharbaoui, Michela Bollati, Giovanni Bertoni, Sandro Recchia, Marcello Marelli, Umberto Piarulli, Sara Pellegrino, Silvia Gazzola
P1319 Peptides stapling as a tool for improving molecular recognition of C-reactive protein
K. Szot-Karpińska, M. Lubos, N. Cwalina, J. Pícha, J. Kaminský, J. Jiráček, M. Majewska, M. Narajczyk, J. Niedziółka-Jönsson
P1320 A structural sight on peptides forming stable LLPS: a combined experimental and computational investigation
Gianfranco Bocchinfuso, Francesca Cavalieri, Daniel O. Cicero, Annachiara Ferrara, Hilda Garay-Perez, Greta Petrella, Marco Savioli, Mariano Venanzi
P1321 From Synthesis To Application: Peptide-Functionalized Nanoformulations For Biomedical Applications
Shubhi Joshi, Gaurav Verma, Avneet Saini
P1322 Sustainable Development of Cross-Linked Films from Rapeseed Meal
Sara Aquilia, Michele Pinna, Anna Maria Papini, Francesco Ciardelli, Luca Rosi, Claudia Bello
P2002 Suspension Bead Loading (SBL): An Economical Protein Delivery Platform to Study URM1's Behavior in Live Cells
Reem Mousa, Abhishek Saha, Yam Alalouf, Pradeep Sadhu, Mahdi Hasan, Shaswati Mandal, Guy Mann, Ashraf Brik
P2003 Synthesis and Conformational Analysis of β,β-Diaryl-α-Amino Acids: on the Way to Molecular Gearing
Timo Zimmer, Armin Geyer
P2004 A green peptide synthesis protocol for improving peptide production: a new approach
Luana Lastella, Barbara Biondi, Ivan Guryanov, Fernando Formaggio, Marco Macis, Antonio Ricci
P2005 A Rapid and Efficient Photochemical Modification Strategy for Late-Stage Peptide and Protein Functionalization
Meng-Qian Zhang, Yan-Mei Li
P2008 Novel Selectively Cleavable Dmb-Based Protecting Groups for Side Chain Amino Groups of Amino Acids
Raimund Maier, Karin Rustler, Thomas Bruckdorfer, Udo Kalejs, Vilnis Liepins, Kai Holland-Nell, Jacqueline Dalski, Stephan Pritz
P2009 Conformational editing and intracellular delivery of intrinsically disordered protein domain
Alexis Jouin, Alastair McEwen, Pierre Poussin-Courmontagne, Bruno Kieffer, Vladimir Torbeev
P2010 Palladium-mediated S-arylation and S-allylation for late stage peptide functionalization and protein conjugation
Ross D. Ballantine, Mingxiang Zhu, Ameni Hadj Mohamed, Aromal Asokan, Linhuan Shen, Mélanie Chenon, Ana Novak, Vincent Robert, Séverine Morisset-Lopez, Massimiliano Beltramo, Carlo Pifferi, Vincent Aucagne, Samir Messaoudi
P2011 Total Chemical Synthesis of Nanobodies for Targeted Cancer Therapy
Keziah Poggenpoel, Kevin Beijerling, William Kelton, Jiwon Hong, Margaret A. Brimble, Iman Kavianinia
P2012 Green Synthesis of a Novel Primary Amine Protecting Group for Solid-phase Peptide Synthesis
Clouie Justin Aguilar, Makhdoom Sarwar, Margaret A. Brimble, Iman Kavianinia
P2013 New Opportunities for Side-Chain Functionalization of Halogenated Amino Acids
Laura Cerveson, Norbert Sewald
P2014 Sculpting cavities in alpha-helical bundles to accommodate ligands of different size and shapes
Katarzyna Ozga, Rokas Petrenas, Joel J. Chubb, Derek N. Woolfson
P2015 Efficient and glycoeconomic synthesis of densely clustered mucin glycopeptides
Arseniy Galashov, Ekaterina Kazakova, Oliver Seitz
P2017 Targeting LAT1 with technetium(I)-99m-labeled tryptophan derivatives for SPECT imaging
Rúben D.M. Silva, Cristina P. Matos, Catarina I.G. Pinto, Joana F. Guerreiro, Lurdes Gano, Filipa Mendes, João D.G. Correia

P2018 Applications of Thiol-ene and Thiol-yne Chemistry for Peptide Stapling and Bioconjugation

Inés Rabadán González, Joshua Mclean, Nikitas Ostrovitsa, Sheila Fitzgerald, Andrea Mezzeta, Lorenzo Guazzelli, Donal O'shea, Eoin M. Scanlan

P2019 Greening Peptide Purification Through Use of Elevated Temperatures

Colin Simpson, Drew Cesta, Sandeep Singh, Jonathan Collins

P2020 Enzymatic Peptide and Protein Bromination: The BromoTrp Tag

Nicolai Montua, Norbert Sewald

P2021 Combining quaternary ammonium compounds and peptides- its application to antimicrobials and beyond

Rolland Lin, Margaret A. Brimble, Brian H. Northrop, Alan J. Cameron

P2022 Fluorescent Amino Acids with large Stokes Shift

Nicolas Schnatmann, Norbert Sewald

P2023 Propargylamine Amino Acids as Constrained Nε-Substituted Lysine Mimetics

Kevin Van holsbeeck, Mathias Elsocht, Steven Ballet

P2024 Synthesis and Structural Characterization of Trifluoromethylated Tetrahydroisoquinolines as New Constrained Aromatic Amino Acid Surrogates

Lilian Loisons, Steven Ballet, Thierry Brigaud, Simon Gonzalez

P2025 Cysteine Thioaldehydes : Photolytic Generation and Reactivity

Ardra Karthika, Matthew Murray, Guillermo Palop, Lee-Ann Keane, Patrick McArdle, Luke Crehan, Nicholas Yates, Alison Parkin, Martin Fascione, Eddie L. Myers

P2026 Application of a new green protocol in Solid-Phase Peptide Synthesis: identification of a new green solvent mixture compatible with ETT/TBEC

G. Vivenzio, M. Scala, P. Campiglia, M. Sala

P2029 Modular Access to Structurally Defined Ubiquitin Chains

Moritz Urschbach, Susanne Huhmann, Luca Ferrari, Dominik Vogl, Dominik Appel, Sascha Martens, Christian F. W. Becker

P2034 Tetrahydropyranyl and Tetrahydrofuranyl as New Backbone Protecting Groups For Enhanced Fmoc SPPS

Samuel J. Paravizzini, Craig A. Hutton, John A. Karas

P2035 Exploring Innovative Liquid-Phase Peptide Synthesis Strategies with Silane-Containing Molecules

Naoki Nishizawa, Akihiro Nagaya, Shota Murase, Yuji Mimori, Michiharu Handa, Ayumu Matsuda, Yutaka Kobayashi, Haruaki Kurasaki, Douglas R. Cary, Keiichi Masuya

P2037 Influence of tryptophan insertion on the fibrillation capacity and antimicrobial efficacy of β-Hairpin Antimicrobial Peptides

Dhanya Mahalakshmi Murali, Jian Xu, Dai Thien Nhan Tram, Pui Lai Rachel Ee

P2040 Cyclic Antimicrobial Peptides Guanidino-based Analogues of Temporin L

I. Boccino, R. Bellavita, M.R. Loffredo, F. Cappiello, B. Casciaro, M.L. Mangoni, P. Grieco, F. Merlino

P2041 From clotting to calcification: Gla-domains as guardians of vascular remodeling

S.M. Agten, A. Gentier, D.P.L. Suylen, S. Peijenborgh, M. Hochstenbach, L.J. Schurgers, T.M. Hackeng

P2042 Design and Synthesis of Peptide Modulators of Gαi/s Proteins

Anna Pepanian, Sabriye Aydoğdu, F. Ayberk Binbay, Pradipta Ghosh, Dehua Pei, Diana Imhof

P2045 Synthesis of Macrocyclic Thiolactone Peptides via Radical Acyl Thiol-Ene

Alby Benny, Eoin M. Scanlan

P2046 Divalent metal ions-directed release of C-terminally modified peptides at 60 °C under mild conditions 102

Rebeca G. Moura, Thiago de Souza Freire, M. Terêsa Machini

P2047 Exploring the Three-Dimensional Structures and Activity Relationships of N-Methyl and N-Desmethyl Analogues of Cyclodepsipeptide Destruxin E

Takayuki Doi, Hiroshi Sato, Hiroshi Nakagawa, Masahito Yoshida

P2048 Green solvents and Safe Reagents in SPPS taking advantage of induction heating and automation: synthesis and HPLC-free purification of difficult sequences

L. Pacini, M. Muthyala, O. Marder, L. Aguiar, R. Zitterbart, P. Rovero, A. Papini

P2049 Simple one-pot peptide cleavage and spontaneous macrocyclization through Paal- Knorr-like reactions using 2,5-dialkylfuran derivatives

Alex Manicardi, Atiruj Theppawong, Marleen Van Troys, Annemieke Madder

P2050 Sustainable Scale-Up of GLP-1 Agonist Peptides through Green Solid Phase Peptide Synthesis

L. Pacini, M. Muthyala, R. Zitterbart, P. Rovero, A. Papini

P2051 Searching for peptide inhibitors of TNFR2/TNFαcomplex formation using a virtual

Paulina Stencel, Artur Giełdoń, Beata Gromadzka, Sylwia Rodziewicz-Motowidło, Marta Spodzieja

P2054 Core-shell type resins for SPPS
Kwanwoo Kim, Seok-In Kim, Yoon-Sik Lee
P2055 Nature-inspired unique antimicrobial peptides from Chilean marine sediments
L. I. Beyer, A.-B. Schäfer, A. Undabarrena, I. Mattsby-Baltzer, D. Tietze, E. Svensson, A. Stubelius, M. Wenzel, B. Camara, A. A. Tietze
P2056 The role of the quorum sensing peptide PlnA in the microbiome-sarcopenia axis
Amélie Descamps, Anton De Spiegeleer, Liesbeth Crombez, Morten Kjos, Evelien Wynendaele
P2058 Unlocking Mango's Peptide Potential: Exploring Bioactive Peptides in Pulp, Peel and Seed
Mahya Bahmani, Michelle L. Colgrave, Cassandra L. Pegg, Angéla Juhász
P2059 Alca1, a cyclotide extracted and isolated from Allexis cauliflora (Violaceae): Synthesis and biological activities
Olivier Ndogo Eteme, Letícia Oliveira Catarin Nunes, Marcela Nunes Argentin, Norival Alves Santos-Filho, Ilana Lopes Baratella Da Cunha Camargo, Eduardo Festozo Vicente
P2060 Microporous Scaffold-Mediated Myeloid Cell Activation in Situ for Robust Peptide-Based Cancer Vaccine
Shao-Hua Zhuo, Wen-Hao Li, Jing-Yun Su, Yan-Mei Li
P2062 The "Sand in a gearbox" effect of antimicrobial peptides: beyond pore formation on model membranes and real bacteria cells
Daniela Roversi, Z. Vaezi, A. Bortolotti, C. Di Stefano, S. Bonacorsi, V. Santucci, C. Aisenbrey, M. M. Konai, Y. Acharya, R. Khosravi-Far, B. Bechinger, J. Haldar, L. Stella
P2064 Design and Structural Analysis of Fluorinated Polyproline-Type Foldamers and their Ability to Interact with Membrane Models
Chloe Cayrou, Simon Gonzalez, Astrid Walrant, Delphine Ravault, Karine Guitot, Sylvie Noinville, Françoise Illien, Sandrine Sagan, Thierry Brigaud, Olivier Lequin, Sandrine Ongeri, Gregory Chaume
P2065 Cyclic peptides as Connexin43 and Pannexin1 inhibitors for the treatment of inflammatory disorders
Debora Iaculli, Arthur Lamouroux, Malaury Tournier, Jade Montgomery, Anne Caufriez, Mathieu Vinken, Filippo Molica, Brenda R. Kwak, Steven Ballet
P2066 Antimicrobial peptidomimetics: design, synthesis, and surface binding
Marc Mathieu
P2067 New Insights Into Antibacterial And Antitumor Effects Of Structurally Versatile Amino Acid- Based And Non-Proteinogenic Compounds
Martina I. Peeva, Maya G. Georgieva, Aneliya A. Balacheva, Svetoslav Dimov, Nikolay T. Tzvetkov
P2068 Chemical Synthesis and Intracellular Delivery of P53 Transactivation Domain (TAD) Constructs
Mo'ath Yousef, Vladimir Torbeev
P2071 Development of a retro-inverso tetrapeptide collagen modulator as anti-aging active principle
Fosca Errante, Marco Pallecchi, Gianluca Bartolucci, Elena Frediani, Francesca Margheri, Lisa Giovannelli, Anna Maria Papini, Paolo Rovero
P2076 A BMP-2-derived peptide conjugated to chitosan improves osteoblast proliferation and antibacterial properties of membranes for oral surgery
Cristian Balducci, Annj Zamuner, Martina Todesco, Andrea Bagno, Antonella Pasquato, Giovanna Iucci, Federica Bertela, Chiara Battocchio, Luca Tortora, Luca Sacchetto, Paola Brun, Erilberto Bressan, Monica Dettin
P2077 Transmembrane peptide-loaded ionic liquid nanocarriers targeting ErbB2-positive cancer
Helal Abujubara, Pankaj Bharmoria, Samantha W. Alvarez, Enoch Appiah, Kasper Moth- Poulsen, Volkan I. Sayin, Alesia A. Tietze
P2078 Characterization of Aurora-A Activation and Allosteric Regulation by TACC-3 Peptidomimetics
Diana Gimenez, Martin Walko, Jennifer A. Miles, Richard Bayliss, Megan H. Wright, Andrew J. Wilson
P2080 Synthesis, conformational analysis and biological activity evaluation of novel antiviral peptides blocking the SARS-CoV-2 cell-entry
M. Quagliata, M.A. Stincarelli, A. Di Santo, L. Pacini, F. Real-Fernandez, R. Arvia, S. Rinaldi, S. Giannecchini, A.M. Papini, P. Rovero
P2083 Synthesis, characterization and biological activity evaluation of a dual-drug based on the antibacterial peptide lugdunin functionalized with the Carbonic Anhydrase sulfonamide inhibitor acetazolamide
M. Quagliata, F. Brenzini Biagioni, S. Giovannuzzi, V. De Luca, C. Capasso, A. Nocentini, C. T. Supuran, P. Rovero, A.M Papini
P2085 Phosphorylation and GlcNAcylation of the RNA polymerase II C-Terminal domain and their effect on recognition by Pin1 and Protein Phosphatase 1
Alina Herfort, Emanuele Piemontese, Corina Maller, Maja Köhn, Oliver Seitz

P2086 How a single non-conventional amino acid can modify the effectiveness of amphibian antimicrobial peptides
Maria Rosa Loffredo, Bruno Casciaro, Giacomo Cappella, Rosa Bellavita, Diego Brancaccio, Floriana Cappiello, Francesco Merlino, Paolo Grieco, Lorenzo Stella, Alfonso Carotenuto, Maria Luisa Mangoni
P2087 Development of bioactive peptides of cosmetic interest through environmentally sustainable processes
L. Sforzi, F. Errante, J. Makowska, P. Rovero, A. Papini
P2088 Thermodynamic and kinetic effects on the activity and selectivity of antimicrobial peptides
F. Carneri, E. Franci, C. Troiano, G. Giaquinto, A. Leoci, D. Roversi, M. R. Loffredo, Y. Park, A. Toto, S. Gianni, M. L. Mangoni, H. Franzyk, L. Stella
P2091 Different approaches in optimisation of miniprotein structure for improvement of affinity towards target and inhibitory activity
Paweł Noceń, Łukasz Berlicki
P2092 Developing autoinducing peptides for regulating the quorum sensing system in S. lugdunensis
Iben Jensen, Benjamin S. Bejder, Martin S. Bojer, Hanne Ingmer, Christian A. Olsen
P2093 Design of fluorinated Aib foldamers: synthesis, structural study and capacity to interact with amyloid proteins
Marine Pietri, Olivier Monasson, Julia Kaffy, Elisa Peroni, Sandrine Ongeri, Grégory Chaume, Nathalie Lensen
P2094 Antimicrobial peptides with varied chemical compositions trigger hyperpolarization of the bacterial membrane
Anasztázia Hetényi, Kaushik Nath Bhaumik, Réka Spohn, Anett Dunai, Lejla Daruka, Csaba Pál, Tamás A. Martinek
P2095 Synthetic glucosylated peptide antigens of the bacterial protein adhesin of nontypeable Haemophilus influenzae to detect antibodies in Multiple Sclerosis
F. Nuti, F. Real-Fernandez, H. Berhane Tesfaya, P. Strauss, R. Aharoni, M. Hurevich, P. Rovero, A.M. Papini
P2098 Design, synthesis, and biological activity of tumor necrosis factor-alpha peptides for adalimumab recognition
Feliciana Real Fernandez, Andrea Di Santo, Kalina Kirilova Kirilova, Silvia Rinaldi, Gabriele Simonini, Paolo Rovero
P2099 Computational design of ultrashort peptide inhibitors for the RBD of the SARS-CoV-2 S-protein
Pengfei Pei, Shi-Zhong Luo
P2100 In vitro effect of the acidic pH on the susceptibility of the antimicrobial peptide Esc(1-21) against respiratory pathogens in cystic fibrosis lung environment
Giacomo Cappella, Maria Rosa Loffredo, Bruno Casciaro, Floriana Cappiello, Arianna Turchini, Maria Luisa Mangoni
P2104 Glycine- and histidine-rich peptides derived from Shepherin I: study of candidacidal activity, metal-chelation properties and mode of action 104
Gustavo Nunes, Lohanna Lopes, Ernani Lacerda, Ana M. C. Ferreira, Carlos Marques, Breno P. Esposito, M. Terêsa Machini
P2107 ReCHEMbinant protein engineering – synthetically enhanced protein therapeutics for Myc targeting
Jan Pascal Kahler, Brecht D. Ellenbroek, Sebastian Pomplun
P2108 Cecropin A-Melittin B Hybrid CA(1-7)M(2-9) Analogs with Improved Antibacterial Activity, Low Toxicity, and Good Protease Stability
Talia Chetty, Jessica T. Mhlongo, Ayman Y. Waddad, Fernando Albericio, Beatriz G. de la Torre
P2109 Optimizing Ham Curing Process: Impact of overripening and the use of nitrifying salts on Peptide Bioactivity.
Noelia Hernández Correas, Cindy Bande-De León, Adela Abellán Guillén, Teresa Muela Clemente, Eva Salazar Serna, Jose María Cayuela García, Luis Tejada Portero
P2110 Functional and structural understanding of the angiotensin II receptor type 1 (AT1R) biased signaling
Margot Hadjadj, Alexandre Murza, Brian Holleran, Yoon Namkung, Stéphane Laporte, Richard Leduc, Mannix Auger-Messier, Pierre-Luc Boudreault
P2111 Molecular determinants of the anti-amyloid function of the cross-interaction of the human cathelicidin LL-37 with islet amyloid polypeptide 106
Valentina Armiento, Hendrik Wunderlich, Aphrodite Kapurniotu
P2114 Use of alternative curing salts in the cured ham production process. Effect on the generation of bioactive peptides
Cindy Bande-De León, Noelia Hernández Correas, Manuel Hernández Lorca, Pablo Gómez San Miguel, Gregorio Molina Valero, Estefanía Bueno Gavilá, María Dolores Gil Martínez, Adela Abellán Guillén
P2116 Development of a potent and specific peptide inhibitor for the E. coli MazF toxin 108
Luiz R. Pizzolato-Cezar, Phelipe Vitale, Carolina D. Lacerda, Mario A. R. Pineda, Roberto K. Salinas, M. Terêsa Machini

P2120 Buforin-Derived Synthetic Peptides with Potential Application in Medicine
Dana Maria Copolovici, Cristian Moisa, Andreea Ioana Lupitu, Oana Gavriliuc, Florina Bojin, Mihaela Dochia, Dorina Chambre, Lucian Copolovici
P2121 Exploring antimicrobial peptides: a new frontier in cancer treatment innovation
Lucian Copolovici, Cristian Moisa, Andreea Ioana Lupitu, Dorina Chambre, Dana Maria Copolovici
P2122 High level soluble expression of human Glucagon-Like Peptide-1 (GLP-1) analogue as a recombinant peptide in E. coli
S. Sushmita, Esha Shukla, Gaurav Pandey
P2124 Combining peptide fragments for the high-throughput generation and screening of large libraries
Anne Zarda, Xingwang Deng, Edward Will, Maylin Romero, Christian Heinis
P2126 Designed Peptide based on Plant Original Cyclic Scaffold facilitate Rheumatoid Arthritis Diagnosis
Mingshu Zhang, Sunithi Gunasekera, Liyan Mei, Erik Jakobsson, Camilla Eriksson, Helena Idborg, Per-Johan Jakobsson, Ulf Göransson
P2127 Highway to Cell: Internalization of CFTR-Stabilizing Peptides as Potent Cystic Fibrosis Treatment
Clementine Teko-Agbo, Emilie Josse, Quentin Seisel, Dean R. Madden, Magali Taulan Cadars, Prisca Boisguerin
P2128 Effects of peptide labeling on fluorescence binding assays to Src homology 2 domains: the case of protein-protein interactions inhibitors of the oncogenic phosphatase SHP2
Sara Bobone, Chiara Fulci, Chiara Innamorati, Alessia Damiani, Claudia Storti, Paolo Calligari, Daniela Roversi, Luca Pannone, Simone Martinelli, Gianfranco Bocchinfuso, Fernando Formaggio, Cristina Peggion, Barbara Biondi, Lorenzo Stella
P2129 Bicyclic peptidomimetics enable efficient transport of proteins across brain endothelium
Maria C. Lucana, Roberta Lucchi, Fabien Gosselet, Cristina Díaz-Perlas, Benjamí Oller-Salvia
P2132 Synthesis of Halogenated Cyclic RGD-Peptides as Potent SMDC-ligands
Beate Nachtigall, Norbert Sewald
P2133 Supramolecular anticoagulants with on-demand reversibility
Millicent Dockerill, Daniel J. Ford, Simona Angerani, Imala Alwis, Luke J. Dowman, Jorge Ripoll-Rozada, Rhyll E. Smythe, Joanna S. T. Liu, Pedro José Barbosa Pereira, Shaun P. Jacksond, Richard J. Payneb, Nicolas Winssinger
P2134 Adalimumab and anti-adalimumab antibodies a novel method of detection and quantification in human sera based on surface plasmon resonance technique
Andrea Di Santo, Fosca Errante, Manuela Capone, Alessandra Vultaggio, Matteo Accinno, Lorenzo Cosmi, Francesco Annunziato, Anna Maria Papini, Paolo Rovero, Feliciana Real-Fernandez
P2136 Innovative liquid chromatographic-mass spectometric technologies to purify and identify biologically active antibody-drug conjugates
Margherita Marino, Paolo Rovero, Walter Mier, Hendrik Rusche, Anna Maria Papini
P2140 High-Throughput Permeability Assay for the Discovery of Cell Penetrating Cyclic Peptides
Hanna Stokes, Noora Azadvari, Parisa Hosseinzadeh
P2143 UL144-derived peptides as modulators of the BTLA/HVEM signaling pathway
Magdalena Lipińska, Piotr Ciura, Sylwia Rodziewicz-Motowidło, Anna Wardowska, Marta Spodzieja
P2144 β-Hairpin peptide fibrillation and bacteria trapping in hydrogel
Peiyan Yu, Mrunmayee Deshpande, Nhan Dai Thien Tram, Pui Lai Rachel Ee
P2145 Peptide Nanonets as Antimicrobial and Anti-inflammatory Agents
Pui Lai Rachel Ee, Nhan Dai Thien Tram, Jian Xu, Dhanya Mahalakshmi Murali, Peiyan Yu, Wai Shiu Fred Wong, Lakshminarayanan Rajamani
P2146 Protein-Protein Interactions: Unraveling the link between HTLV-1 Tax-1 oncoprotein and human Drosophila discs large tumor suppressor
Wael El Yazidi Mouloud, Sibusiso B. Maseko, Yasmine Brammerloo, Inge Van Molle, Adria Sogues Castrejon, Charlotte Martin, Remaut Han, Steven Ballet, Oleksandr Volkov, Jean-Claude Twizere
P2148 Esc peptides for novel therapeutics against Pseudomonas aeruginosa pulmonary infections: beyond antimicrobial activity
Maria Rosa Loffredo, Loretta Ferrera, Floriana Cappiello, Bruno Casciaro, Giacomo Cappella, Ivana D'Angelo, Y. Peter Di, Maria Luisa Mangoni
P2150 Synthesis and antimicrobial activity of hybrid peptides as dual inhibitor of intracellular targets in ESKAPE pathogens
Aman Kumar Mahto, Rikeshwer Prasad Dewangan

P2152 Disrupting the RANK-TRAF6 Interface: A Novel Approach for Bone Metastases Treatment
Sofia A. Martins, Cátia Rosa, Marco Cavaco, Rita Melo, David Andreu, Javier Valle, Vera Neves, Miguel A. R. B. Castanho, João D.G. Correia
P2154 Targeting MRGPRX2 with venom peptide P17: A therapeutic avenue for DSS-Induced colitis
Karthi Duraisamy, Mukesh Kumar, Abdullah Nawabjan, Emily Kwun Kwan Lo, Benjamin Lefranc, Elsa Bonnafé, Michel Treilhou, Hani El-Nezami, Billy K.C. Chow, Jérôme Leprince
P2158 Fight-TB: Design of Novel Vaccine Antigens to Fight Tuberculosis
Marina Sala, Giovanni Barra, Maria Carmina Scala, Pietro Campiglia, Rita Berisio, Alessia Ruggiero
P2161 Advances on the green synthesis of Etelcalcetide
Koula Achilleos, Andreas Sergides, Styliana Mirallai, Charalambos Ioannou, Vicky Nicolaidou, Christos Petrou, Christakis Sergides, Yiannis Sarigiannis
P2163 Determination of the Oligomerization States of the Transmembrane Domains from SARS-CoV-2 Membrane Proteins
Tyler G. Hobart, Andrew M. Rehak, Timothy M. Reichart
P2164 Exploiting the features of short peptides to recognize specific cell surface markers
Michela Buonocore, Angelo Santoro, Enza Napolitano, Elena Ciaglia, Francesco Montella, Francesca Boccafoschi, Valeria Perugini, Matteo Santin, Anna Maria D'Ursi
P2167 Obtaining, Optimizing, and Evaluating the Activity of Anticancer LfcinB-Derived Peptides on in vitro Models against Melanoma
Karla Rodriguez, Jose Mendoza, Javier Garcia, Zuly Rivera, Joel Lopez, Ricardo Fierro
P2169 Design and characterization of a multistage peptide-based vaccine platform to target Mycobacterium tuberculosis infection
Chiara Bellini, Emil Vergara, Fruzsina Bencs, Szilvia Bősze, Denis Krivić, Bernadett Bacsa, Sára Eszter Surguta, József Tóvári, Rajko Reljic, Kata Horvati
P2171 Engineering mRNA display for GPCR drug discovery – de novo discovery of cyclic peptide GLP1R binders
Esther Meron, Jens C. Nielsen, Mads M. Nygaard, Dennis Worm, Ulrike Leurs, Morten Lundh, Joseph M. Rogers
P2172 Novel functionalized acylation reagents for Glyand Lys-His tag acylation on proteins
Delphine N. Møller, Christian Kofoed, Mikkel B. Thygesen, Kasper K. Sørensen, Tuule Treiberg, Knud J. Jensen
P2173 KLVFF peptide functionalization of graphene oxide as suitable Aβ biomarker detection platform
Rita Turnaturi, Giuseppina Sabatino, Giuseppe Di Natale, Giuseppe Nocito, Angela Scala, Viviana Scuderi, Simona Filice, Antonino Mazzaglia, Silvia Scalese, Giuseppe Pappalardo
P2174 On-Resin Synthesis and Modification of Novel Peptide Heterocycles
Ciaran McGrory, Eddie Myers
P2175 Use of Green Solid Phase Peptide Synthesis to Obtain Tripeptide-85
M. C. O. A. Brasil, C. R. Princival, V. Kuchenbecker, E. Katekawa, W.V. Magalhães, E.M. Cilli
P2176 Affinity Peptide Extraction Using Small-Molecule, Thermoresponsive Zwitterionic Liquid Materials
Yen-Ho Chu, Yu-Hsin Chung
P2178 Optimization of pDNA delivery by WRAP cell-penetrating-peptide
K. Konate, S. Deshayes, P. Boisguérin, E. Vivès, S. Faure, P. De Santa Barbara
P2180 Application of coumarin-type chemiluminophore toward in vivo detection
Yukie Nohara, Keiko Taniguchi, Hiromi Ii, Shun Masuda, Hiroko Kawakami, Masakatsu Matsumoto, Susumu Nakata, Toshiyuki Sakai, Taku Yoshiya
P2181 Development of novel automated solid phase peptide synthesizers ensuring economic, effective, scalable and green synthesis
Krisztina Kiss, Soma Ránky, László Molnár
P2183 Discovery and Characterization of a Short Peptide Possessing a Latent Warhead Undergoing Intramolecular Cyclization in the Aqueous Environment
Riku Katsuki, Masumi Taki
P2186 A fishing technology of glycated peptides
F. Nuti, M. Marino, P. Ledwon, R. Latajka, A. Lapolla, M. Chorev, P. Rovero, A.M. Papini
P2188 Programmable Chemical Synthesis & in-situ Modifications of Peptides
Jacopo Zero, Tristan Tyler, Leroy Cronin
P2191 Peptide-based Coacervates as Evolutionary Roads of Catalytic Function
David Reis, Sara Pereira, Ana P. Ramos, Pedro M. Pereira, Leonor Morgado, Ana S. Pina
P2194 An all-in-one automated peptide purification and synthesis solution
Luísa Aguiar, M. Muthyala, S. Lüdtke, Dominik Sarma, Robert Zitterbart

P2195 Riding new Tides: Synthesis of oligonucleotides on PurePep® Chorus
Luisa Aguiar, Dominik Sarma, Robert Zitterbart
P2196 The first endocrine-disrupting chemical inhibiting a coactivator peptide's binding
Masaki Iwamoto, Takahiro Masuya, Mari Hosose, Tomaka Ishibashi, Ayami Matsushima
P2197 PYY peptides and Fc-PYY conjugates selectively agonise Y2R
Elise Bernard, Siobhan O'Brien, Nicole Burmeister, Jacqueline Naylor, Stephanie Oldham, Arthur Suckow, Hamid Salari, Maria A. Bednarek
P2200 Pipeline for development of acylated peptide-based CGRP receptor antagonist with extended half-life for migraine treatment
Jens Bjelke Kristensen, Lisbeth Elster, Morten Lundh, Borja Ballarín-González, Flora Alexopoulou, Martin Kræmer, Ditte Marie Jensen, Ulrike Leurs, Jens Christian Nielsen, Henrik H. Hansen, Matilda Degn, Kristian A. Haanes
P2201 Peptide-conjugates derived from bacterial ParE toxin: synthesis and biological studies
Jose Gregorio Martín Bedoya, Naiara Maia Pereira, Reinaldo Marchetto
P2202 Identification and Characterization of Small Molecule Nucleotide Binding and Oligomerization Domain 2 (NOD2) Receptor Agonists
Liora E. Wittle, Karl L. Ocius, Marcos M. Pires
P2203 Influence of Hydrophobic Moieties on the Formation and Stability of Collagen Heterotrimers
Maša Stopinšek, Valdrin Islami, Giuseppe Antoniazzi, Philipp Bittner, Renato Zenobi, Helma Wennemers
P2207 Exploring Novel Peptide-Based Natural Products as a Payload for the Generation of Antibody and Peptide-Assisted Targeted Cancer Therapy
Suzie M. Chan, Makhdoom Sarwar, Jeanine Giarolla Vargas, J. da Silva, Margaret A. Brimble, Iman Kavianinia
P2208 Bispecific DNA-Peptide Conjugates for Targeting Cancer Cells
Pritam Ghosh, Huyen Dinh, Oliver Seitz
P2209 Memantine and Adamantine KLVFF Conjugates: Synthesis, Ab interaction and Effects on Recognition Memory in Mice.
Giuseppe Pappalardo, Eleonora Bocchieri, Stefania Zimbone, Maria Laura Giuffrida, Giuseppe Di Natale, Giuseppina Sabatino, Graziella Vecchio, Santina Chiechio
P2210 Chemoselective arylation of amino acids, peptides and proteins
Ahmed M. Embaby, Sanne Schoffelen, Christian Kofoed, Morten Meldal, Frederik Diness
P2211 Glycopeptides as synthetic antigenic probes for MOG Antibody-Associated Disease
Silvia Braccia, Feliciana Real-Fernandez, Christine Patte-Mensah, Christian Klein, Nicolas Collongues, Jérôme de Sèzed, Paolo Rovero, Anna Maria Papini
P2212 Designing self-assembling lipopeptide for tissue regeneration
Dhriti Shenoy, Lauren Roberts, Michela Pesenti, Kaliroi Peqini, Stefania Crespi, Cécile Echalier, Gilles Subra, Sara Pellegrino
P2213 Optimizing biological activity of Lactoferricin derived peptides: Conjugation with non peptidic motifs as a strategy
Daniel Castellar, Javier García, Zuly Rivera, Ricardo Fierro
P2215 Introducing Novel Titanium Binding Peptides; Effect of the Sequence of Amino Acids on Their Adsorption Energy to TiO2
Roja Rahmani, Alexander P. Lyubartsev
P2216 Peptides Conjugated to Sugars as Therapeutics for Ebola & Cancer disease
Fayna García-Martín, Óscar Suárez, Gonzalo Millán, Filipa Marcelo, Ana Gimeno, Elena Lalinde, Jesús R. Berenguer, Juan Anguita, Héctor Busto, Jesús Manuel Peregrina, Asuka Nanbo, Jesús Jiménez-Barbero, Francisco Corzana
P2217 Synthetic strategies to prepare novel bioactive lysine and peptide conjugates with triazolium derivatives
Claudia Bello, Michał Jewginski, Patrycja Ledwon, Francesca Nuti, Paolo Rovero, Rafal Latajka, Anna Maria Papini
P2222 Effect Of Azaglycine On The Internalization Of Penetratin Derivatives
Karima Tarchoun, Dóra Soltész, Ildikó Szabó, Zoltán Bánóczi
P2223 Molecular Design for Cytosolic Delivery of Antibodies Using an Intracellular Delivery Peptide
Yoshimasa Kawaguchi, Shiroh Futaki
P2224 Synthesis and characterization of nucleolin receptor recognizing F3 peptide fragment-drug conjugates
Bettina Basa, Diána Vári-Mező, Kata Nóra Enyedi, József Tóvári, Szilvia Bősze, Krisztina Kiss, Gábor Mező
P2225 Leuprorelin Peptide Drug Complexes with Permeability Enhancers to Increase Bioavailability
Adel Penhasi, Zeina Zaatari, Moustafa Ghanyem, Deborah E. Shalev

P2267 Application of Deep learning Imputation to Peptide Bioactivity and Property Prediction
Samar Mahmoud, Himani Tandon, Matthew Segall, Ruibo Zhang, Robert E Thompson, Jeffrey W Schubert, Dipannita Kalyani, Dan Sindhikara
P2268 PICKAPEP: Visualization and Calculation of Cyclized and Modified Peptidomimetics Made Easy
Vanessa Erckes, Mattis Hilleke, Clemens Isert, Christian Steuer
P2274 Metalloproteins Investigation for the Development of new Inhibitors
Matteo Orlandi, Piero Procacci, Marina Macchiagodena, Fabrizio Carta, Claudiu T. Supuran, Marco Pagliai
P2275 Deep learning for Peptide Classification: From Data to Effective Models
Marko Njirjak, Marko Babić, Patrizia Jankovic, Erik Otović, Daniela Kalafatovic, Goran Mauša
P2276 Scoring the Synthesizability of Any Novel Non-natural Amino Acid
Gökçe Geylan, Mikhail Kabeshov, Samuel Genheden, Thierry Kogej, Ola Engkvist, Leonardo De Maria
P2277 Chemoenzymatic synthesis of protein pseudo-[2]rotaxane
Elodie Soehnlen, Shalene Ofner, Judit Osz, Natacha Rochel, Vladimir Torbeev
P2279 Structural and Mechanistic Insights Into the Advances of FXIIIa Inhibitors
Ivan P. Bogdanov, Arijit Biswas, Diana Imhof, Nikolay T. Tzvetkov
P2280 Can large scale structural modeling be applied to peptide screening and deorphanization?
Shaima Hashem, Jérôme Leprince, Pierre Tuffery
P2281 Impact of backbone N-amination on the stability of parallel beta-sheets
Syrah Starnes, Seth Horne, Juan Del Valle
P2283 How do Disulfide, Thioacetal and Lanthionine Bridges Influence the Conformations of Macrocyclic Peptides?
William T. P. Darling, Lianne H. E. Wieske, Declan T. Cook, Abil E. Aliev, Laurent Caron, Emily J. Humphrys, Angelo Miguel Figueiredo, D. Flemming Hansen, Máté Erdélyi, Alethea B. Tabor
P2284 Characterizing bradykinin-derived membrane-sensing peptides through Molecular Dynamics
Carlo Guardiani, A. Giacomello
P2286 Structural Ensemble of Peptide-induced Membrane Poration by Heterogeneous Magainin Peptides
Bing Zan, Charles Chen, Martin Ulmschneider, Jakob Ulmschneider-Mayer
P2288 Helical peptides bearing dual triplet-spin/fluorescent labels enable the first parallel distance measurements by electron spin resonance and Förster resonance energy transfer
Marta De Zotti, Arnau Bertran, Chiara Dalla Torre, Jack Sawyer, Derren Heyes, Sam Hay, Christiane Timmel, Marilena DI Valentin, Alice Bowen
P2289 Antiparallel β-Double Helices from α,γ-Hybrid Peptides
DRGKoppalu R. Puneeth Kumar, Gijo George, Uddipan Bhattacharya, Arnab Mukherjee, Hosahudya N. Gopi
P2290 Aromatic-aromatic interactions as an intrinsic modulator of amyloid formation
Artem Pavlov, Beate Koksch
P2291 Structure-Activity Relationship Study of β-Hairpin Peptides with Antibiotic Activity
Aparna Palakkurussi Rathessan, Suvrat Chowdhary, Johannes Kupke, Marcus Fulde, Beate Koksch
P2293 Molecular insights into the lipid bilayer interaction of the cyclic lipodepsipeptide tolaasin
Benjámin Kovács, Tünde Juhász, Niels Geudens, José C. Martins, Tamás Beke-Somfai
P2294 A comprehensive structural study on AT1 Receptor Variants: Towards the Design of Potent Biased Agonists
Malihe Hassanzadeh, Audrey Collette, Brian Holleran, Stéphane Laporte, Richard Leduc, Pierre-Luc Boudreault
P2295 Structural analysis of a peptide-based simplified model reproducing SARS-CoV-2 S RBD/ACE2 binding site
Michela Buonocore, Rosanna Lucignano, Angelo Santoro, Anna Maria D'Ursi, Delia Picone
P2296 Planar aggregation of the influenza viral fusion peptide alters membrane structure and promotes poration
Richard W. Pastor, Amy Rice, Paul S. Blank, Joshua Zimmerberg
P2298 Unlocking HIV's Secrets: Synthesis of Histone H3K36Me3 to Investigate LEDGF Protein Binding to Modified Nucleosomes using Cryo-Electron Microscopy
Ramesh M. Chingle, Samanthreddy Kedika, Tao Jing, Dmitry Lyumkis, Terrence R. Burke
P2300 Elucidating the role of phosphoserines in Tau protein by combined computational and synthetic methods
Martina Tuberti, Fintan Kelleher, Gemma K. Kinsella
P2301 Theoretical and biological studies of oxytocin containing conformationally fixed analogue of Tyrosine Tic(OH) (1,2,3,4-Tetrahydro-7-hydroxyisoquinoline-3-carboxylic acid)
Michal Lebl, Jiří Jiráček, Marta Lubos, Jakub Kaminský
P2305 Can we improve Coarse-Grained simulations of peptide selfassembly and aggregation by cluster analysis inputs?
Marko Babić, Daniela Kalafatović

P2306 Design and Exploration of Tubular Assemblies of Hybrid Cyclic Peptides
Souvik Panda Mahapatra, Hosahudya N. Gopi
P2307 Antimicrobial and cell-adhesive peptide-based hydrogels for biomedical applications
Nabila Mehwish, Patricia López-Gómez, Maria Pau-Ginebra, Carles Mas-Moruno
P2309 De novo design of collagen-like peptides for cell-based technologies
Maryana Baran, Eileen Gentleman, Maxim Ryadnov
P2310 Self-assembling dipeptides as vehicles for the delivery of peptide therapeutics
Peter Divanach, Eirini Fanouraki, Vagelis Harmandaris, Anastassia N. Rissanou, Anna Mitraki
P2314 Self-assembly of histidine-containing peptides and applications
C.P Apostolidou, Chrysoula Kokotidou, Varvara Platania, Vasilis Nikolaou, Georgios Landrou, Emmanouil Nikoloudakis, Georgios Charalambidis, Maria Chatzinikolaidou, Athanassios G. Coutsolelos, A. Mitraki
P2315 Tuning the Assembly of Proteins and Peptides in Lipid Bilayers
Sebastián Jiménez Millán, Ivan Sasselli, Armando Maestro
P2318 Design of fluorinated peptide-based hydrogels for sustained drug release application
Aurélie Honfroy, Jolien Bertouille, Ana-Maria Turea, Thibault Cauwenbergh, Jessica Bridoux, Nathalie Lensen, Jessica Mangialetto, Niko Van den Brande, Jacinta F. White, James Gardiner, Thierry Brigaud, Steven Ballet, Sophie Hernot, Grégory Chaume, Ch
P2319 Single glycans shape viscoelastic properties of a mucin-inspired peptide hydrogel
Jonas Proksch, Marlene Dal Colle, Frederick Heinz, Jacqueline Gottwald, Robert Schmidt, Martina Delbianco, Bettina Keller, Ulrike Alexiev, Michael Gradzielski, Beate Koksch
P2330 Cardiolipin-specific peptides, designed to possess cell penetrating ability, rescue mitochondrial dysfunction
Gwangsu Shin, Dongwoo Kim, Yoonhwa Choi, Hyeong woon Choe, Seung Hee Yang, Jaehoon Yu
P2331 Efficient synthesis of three-helix bundle proteins using ligase-mediated trimerization of HIV-1 fusion inhibitor T20 on a dendrimer scaffold
Xiaohong Zhang, Yiyin Xia, Chuan-Fa Liu, James P. Tam
P3332 Broaden the Applications of Peptidyl Asparaginyl Ligases
Yiyin Xia, Hongxia Li, Jiayong Liu, James P.Tam, Chuan Fa Liu

LECTURES

Proceedings of the 37th European Peptide Symposium
Michal Lebl, Editor
European Peptide Society, 2024 http://doi.org/10.17952/37EPS.2024.L003

Synthesis of β-turn mimics without stereogenic carbons: expanding the modular approach for diversity-oriented synthesis of tetrasubstituted benzotriazepinones

Xiaozheng Wei,[1] Sarah-Maude Harvey, [1] Debora Iaculli,[1,2] Steven Ballet,[2] and William D. Lubell[1]

[1]Département de Chimie, Université de Montréal, 1375 Ave. Thérèse-Lavoie-Roux, Montréal, Québec, H2V 0B3, Canada; [2]Research Group of Organic Chemistry, Departments of Chemistry and Bioengineering Sciences, Vrije Universiteit Brussel, 1050 Brussels, Belgium

Introduction

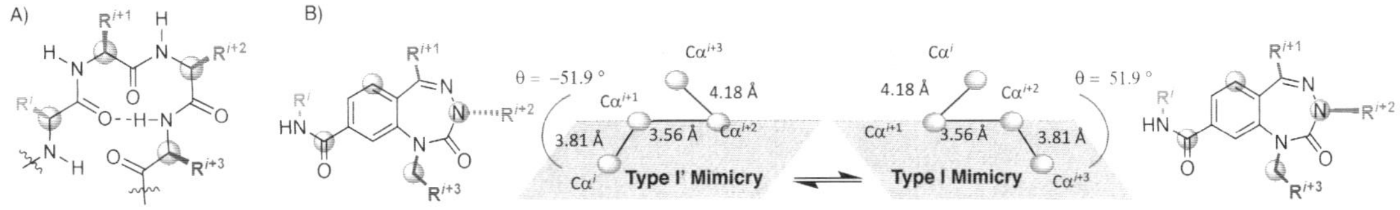

Figure 1. A) β-turn indicating α-carbons, B) dynamic turn mimicry by benzotriazepinone

Benzotriazepinones are "privileged structures" exhibiting promising biological and medicinal properties [1]. Benzotriazepinone crystal structures have shown that the N3 nitrogen is pyramidalized and exists in dynamic equilibrium between two configurations [2,3]. In the context of peptide science, X-ray analysis indicated tetrasubstituted benzotriazepinones replicate the topology of β-turn secondary structures [2]. Superimposing key atoms of the tetrasubstituted benzotriazepinone with the four α carbon atoms ($C\alpha^{i}$–$C\alpha^{i+1}$–$C\alpha^{i+2}$–$C\alpha^{i+3}$) within a tetrapeptide has demonstrated capacity to mimic both type I and I' β-turns which are topologically defined by the dihedral angle θ (Figure 1) [2,4].

A modular synthesis route to tetrasubstituted benzotriazepinones was developed starting from *tert*-butyl 4-acetyl-3-aminobenzoate (Figure 2) [2]. Aldol condensation provides the *i*+1 residue substituent. Annulation with an aza-glycine residue adds the *i*+2 substituent. The *i*+3 and *i* residue substituents are introduced by amide coupling and alkylation.

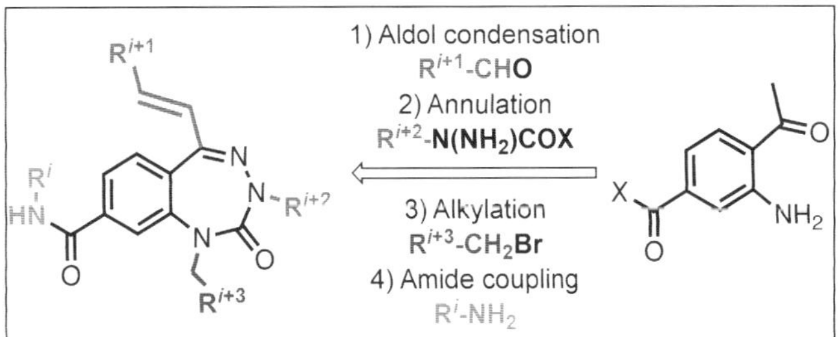

Figure 2. Modular benzotriazepinone turn mimic synthesis

A tetrasubstituted benzotriazepine library was rationally designed, synthesized, and shown to contain urotensin receptor (UT) allosteric modulators [2,5]. A G protein-coupled receptor, UT has two endogenous ligands, *h*UII and URP [5]. Differentiation of pathophysiological roles of *h*UII and URP has been pursued for medical applications, such as cardiovascular disease therapy. Tetrasubstituted benzotriazepinones exhibited respectively selective modulator ability on the activity of *h*UII and URP as positive and negative allosteric modulators with neutral activity on the ligand counterpart [5].

Results and Discussion

Diversity-oriented methods were studied to expand the modular route to β-turn *i*+1 and *i*+3 substituents by the synthesis of amino acid **1** and benzophenone analog **2** (Scheme).

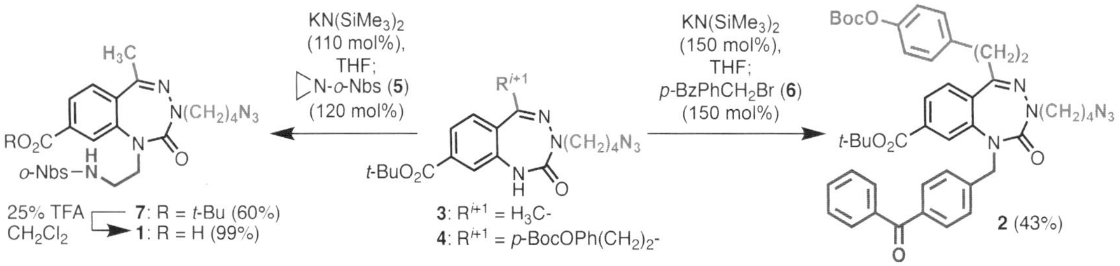

Scheme. Synthesis of tetrasubstituted benzotriazepines **1** and **2**

Benzotriazepines **3** and **4** were respectively prepared from 4-keto-3-aminobenzoates using the reported annulation protocol [2]. The aza-glycinate derived from *p*-nitrophenyl chloroformate and benzophenone hydrazone reacted with 3-aminobenzoates, *i*-Pr_2EtN in CH_2Cl_2 to give glycinamides. Benzotriazepines **3** and **4** were obtained by glycinamide alkylation with $Br(CH_2)_4Cl$ and Et_4NOH in THF, annulation with 1N HCl, and chloride displacement with NaN_3 in 4:1 NMP:H_2O [2].

Respective alkylations of benzotriazepinones **3** and **4** using $KN(SiMe_3)_2$ in THF with *N-o*-nitrosulfonyl (*o*-Nbs)aziridine (**5**) [6] and *p*-benzoyl benzyl bromide (**6**) afforded *o*-Nbs-amino ethyl and benzophenone benzotriazepinones **7** (m/z = 601.2) and **2** (m/z = 771.5). Subsequent treatment of *o*-Nbs-amino ester **7** with F_3CCO_2H in CH_2Cl_2 gave acid **1**, which exhibited diastereotopic methylene proton signals at 3.41 and 4.30 ppm indicative of a chiral nitrogen center.

Modular synthesis of β-turn benzotriazepinone mimics has been expanded through diversification of the *i*+3 position substituent by novel alkylation chemistry which has respectively provided amino acid **1** and benzophenone **2** for applications in peptide synthesis and photoaffinity labeling.

Acknowledgments

We are grateful for funding from the Natural Sciences and Engineering Research Council of Canada (NSERC, Discovery Research Project RGPIN-2019-04079), the Canadian Institutes of Health Research (CIHR, Project PJT-168839), and the Fonds de Recherche du Québec - Nature et Technologie (FRQNT, CGCC, FRQNT-2020-RS4-265155-CCVC). Assistance is acknowledged from members of the Université de Montréal facilities: Dr. P. Aguiar (NMR spectroscopy) and Dr. A. Fürtös (mass spectrometry). D.I. and S.B. thank the Research Foundation Flanders (FWO Vlaanderen) and the Research Council of the VUB for support of Strategic Research Programmes SRP50 and SRP95. S.M.H. thanks NSERC and FRQNT for Summer undergraduate research fellowship support.

References

1. Harvey, S.-M.; Wei, X.; Darince, D. and Lubell, W. D. Benzotriazepine synthesis, conformational analysis, and biological properties. *Can. J. Chem.* **2024**, *102*, 469-498. https://doi.org/10.1139/cjc-2023-0146
2. Wei, X.; Douchez, A.; Lubell, W. D. 1,3,5,8-Tetrasubstituted 1,3,4-benzotriazepin-2-one scaffolds for β-turn mimicry without Stereogenic Carbon Centers: synthesis and conformational analysis. *J. Org. Chem.* **2023**, *88*, 4633–4648. https://doi.org/10.1021/acs.joc.3c00113
3. Douchez, A. and Lubell, W. D. Chemoselective Alkylation for Diversity-Oriented Synthesis of 1,3,4-Benzotriazepin-2-ones and Pyrrolo[1,2][1,3,4]benzotriazepin-6-ones, Potential Turn Surrogates *Org. Lett.*, **2015**, *17*, 6046-6049. https://doi.org/10.1021/acs.orglett.5b03046
4. Hata, M. and Marshall, G. R. Do benzodiazepines mimic reverse-turn structures? *J. Comput. Aided Mol. Des.* **2006,** *20*, 321-31. https://doi.org/10.1007/s10822-006-9059-x
5. Wei, X.; Diarra, S.; Douchez, A.; Dallagnol, J. C. C.; Hébert, T. E.; Chatenet, D. and Lubell. W. D. Urotensin II Receptor Modulation with 1,3,4-Benzotriazepin-2-one Tetrapeptide Mimics. *J. Med. Chem.* **2023**, *66*, 14241-14262. https://doi.org/10.1021/acs.jmedchem.3c01307
6. Sanz-Rodríguez, C.E., Hoffman, B., Guyett, P.J., Purmal, A., Singh, B., Pollastri, M.P. and Mensa-Wilmot, K. Physiologic targets and modes of action for CBL0137, a lead for human african trypanosomiasis drug development. *Mol. Pharmacol.* **2022**, *102*, 1-16. https://doi.org/10.1124/molpharm.121.000430

https://doi.org/10.17952/37EPS.2024.L005

Fundamental aspects of SPPS and the Green future

Stephen Kent

Department of Chemistry

THE UNIVERSITY OF
CHICAGO

European Peptide Symposium
Florence, Italy
August 2024

In 1963, Merrifield introduced solid phase peptide synthesis.

After initial fierce skepticism, SPPS has become almost universally used for the chemical synthesis of peptides, in both academic research and in industry.

In this brief talk, I will describe the fundamental physicochemical properties of peptide-resins that make SPPS so useful.

I will suggest ways to retain that utility in a 'Green' SPPS.

Then I will explain how to combine Green SPPS with chemical ligation for more efficient production of peptide molecules.

Solid Phase Peptide Synthesis (Merrifield 1963)

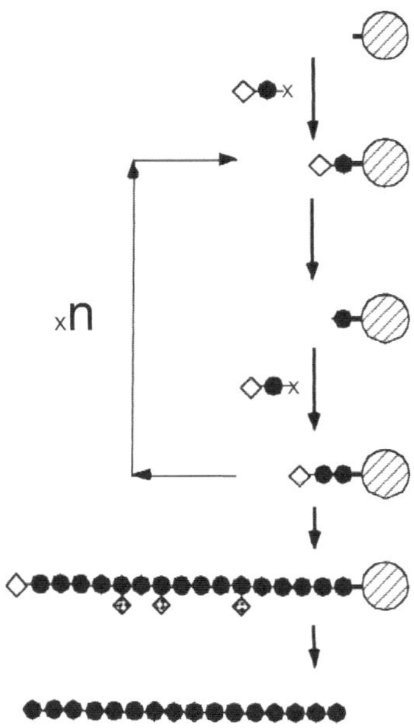

- C-terminal Xaa attached to resin beads
- stepwise chain elongation
- maximal protection of side chains
- global deprotection/release from resin

Is SPPS ‘Green’?

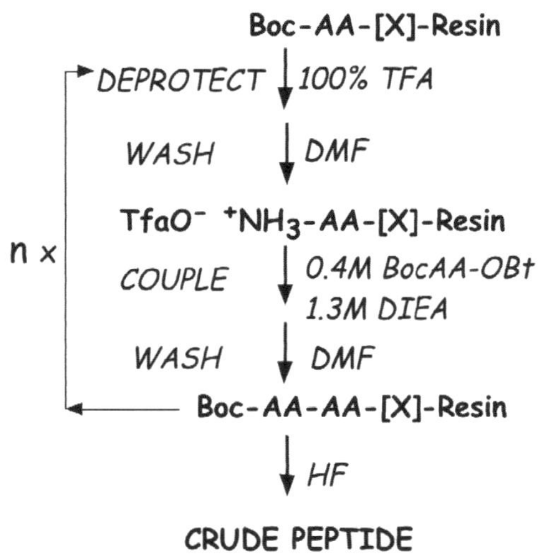

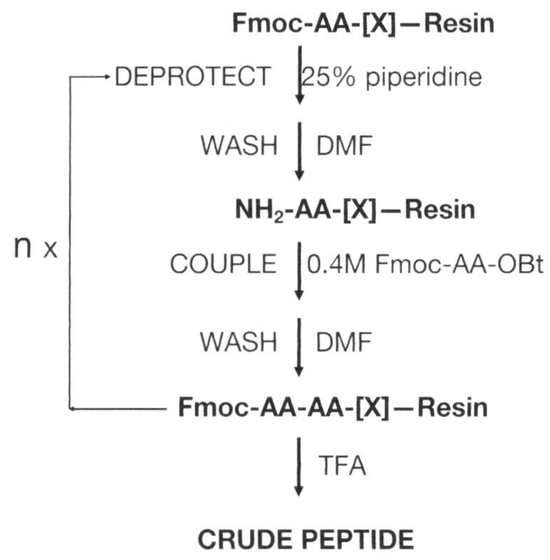

Is SPPS 'Green'?

Boc-AA-[X]-Resin

NO!

Fmoc-AA-[X]—Resin

Poor atom economy:

* maximal use of protecting groups
* large excesses of reactants

Process mass intensive (PMI):

* corrosive/toxic chemicals
* inefficient purification
* high volume waste streams

Can we preserve the inherent advantages of SPPS for 'Green' peptide synthesis?

Advantages of SPPS

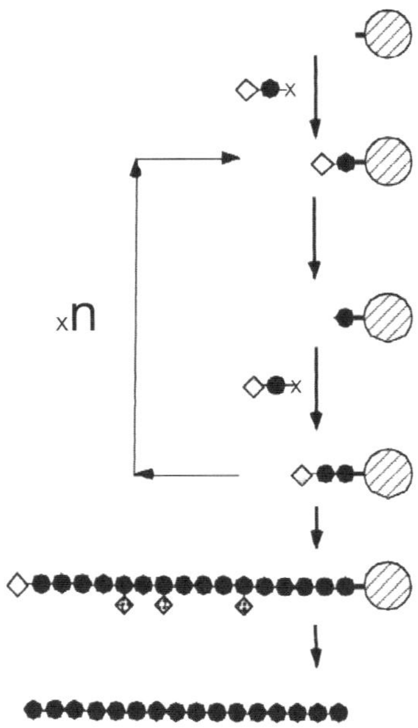

- ***enhanced solvation***
- stepwise synthesis
- purification by filtration
- quantitative recoveries
- general protocols
- automation

Principles of SPPS Polymer Chemistry

(Paul Flory)

resin beads:

- randomly crosslinked
- IPN

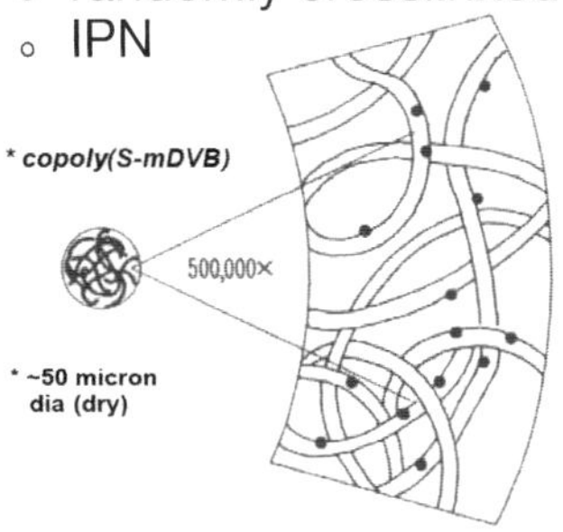

synthesis occurs ***within*** solvent-swollen beads:

~10^{14} peptides per bead

resin-bound peptides are effectively in solution - $\tau_c < 10^{-8}$ sec

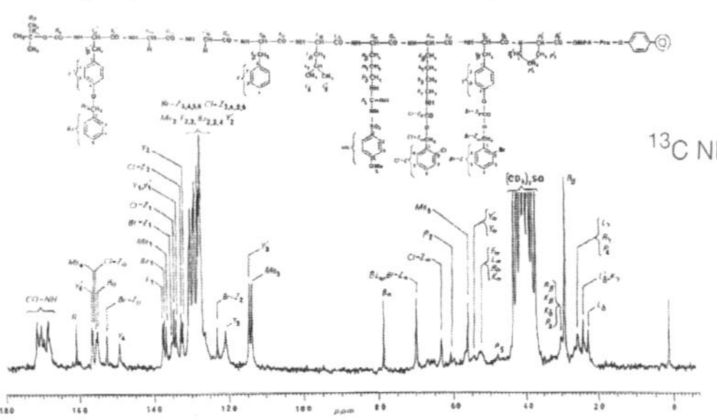

^{13}C NMR peptide-resin

peptide-resin super swelling

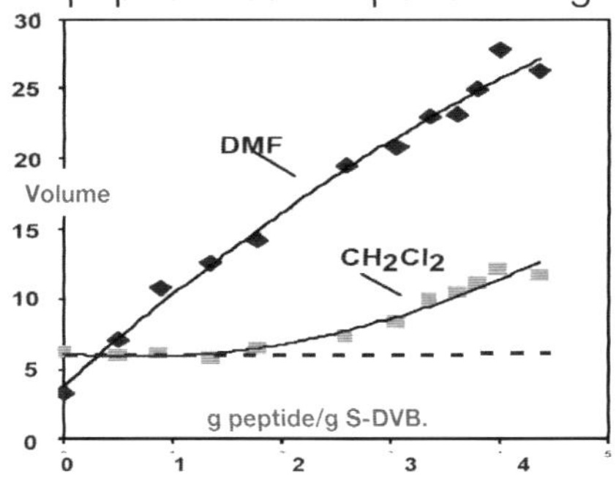

Solvation of resin-bound peptides

Enhanced solvation of peptides within solvent-swollen resin beads is the *fundamental reason* for the *efficacy* and *versatility* of SPPS

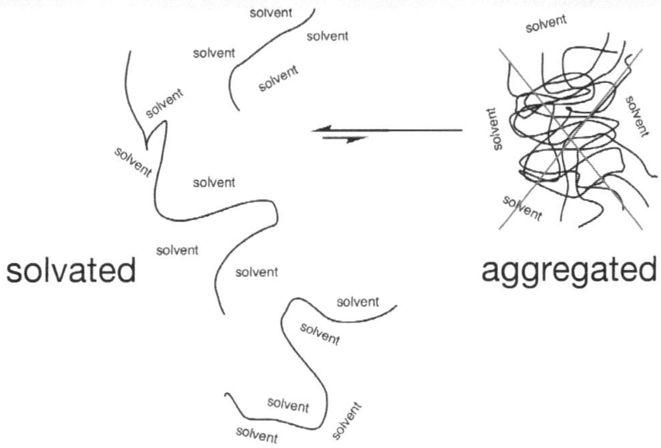

disfavored in peptide-resin

- *cross-links* prevent phase separation
- *dissimilar* properties of resin & peptide

the extent of solvation depends on the *relative free energies* of these two states

Dang, Dhayalan, & Kent
Org. Lett. 2015, **17**, 3521–3523

Features of a Green SPPS

- atom economy
  - minimal protecting groups
  - minimal excess reactants
- benign reactants & reagents
- benign solvent
- efficient purification (PMI)

"Controlled Synthesis of Peptides in Aqueous Solution.

1. The Use of α-Amino Acid N-Carboxyanhydrides." (NCAs)

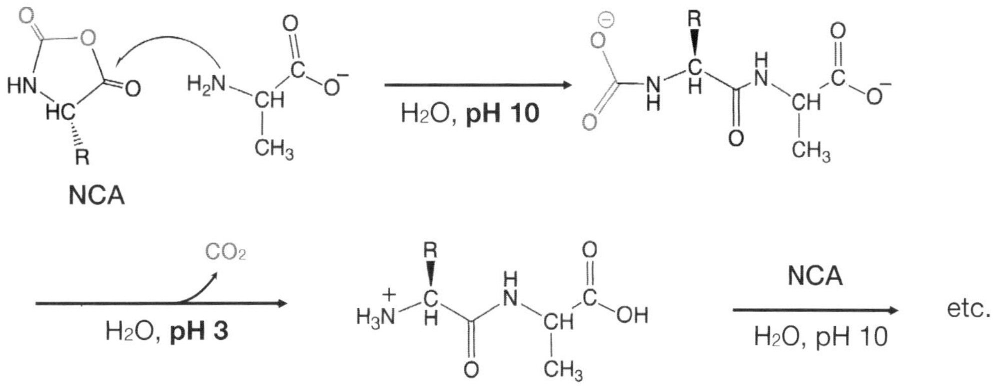

- **atom economy** ✓
  - minimal protecting groups
  - minimal excess reactants
- **benign reagents** ✓
- **benign solvent** ✓

Denkewalter, et al.
J.Am.Chem.Soc.
88, 3163-3164 (1966)

SPPS using α-Amino Acid N-Carboxyanhydrides

on ChemMatrix resin

NCA

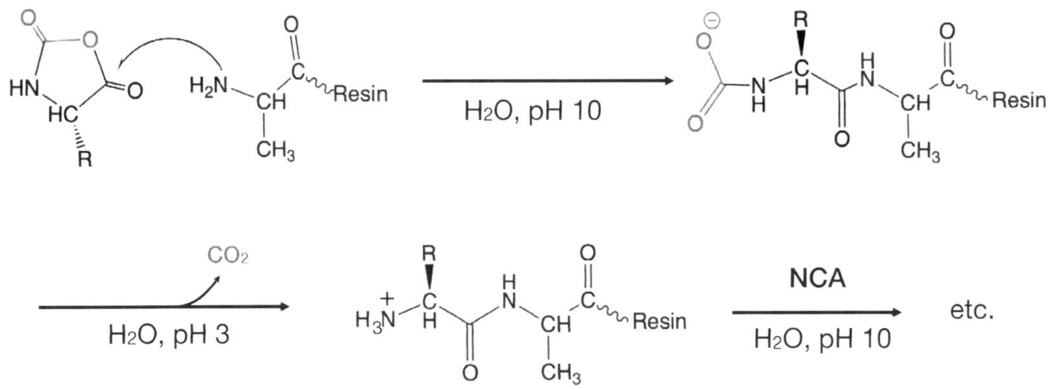

De Marco, Gentilucci, et al., *ACS Sustainable Chem. Eng.* **1,** 566–569 (2013)

Green Solid Phase Peptide Synthesis

Atom economy

- N-carboxyanhydrides
  - minimal side-chain protection
    - Lys
    - Ser/Thr(?)
- high resin loading
  - >1 millimol/g resin
- maximum ***concentration*** of NCA
- minimal ***excess*** of NCA

Solvent

- water

Resin

- aqueous solvent-compatible

Preparation of chemically defined resins for SPPS

BioBeads Sx1 (Biorad)

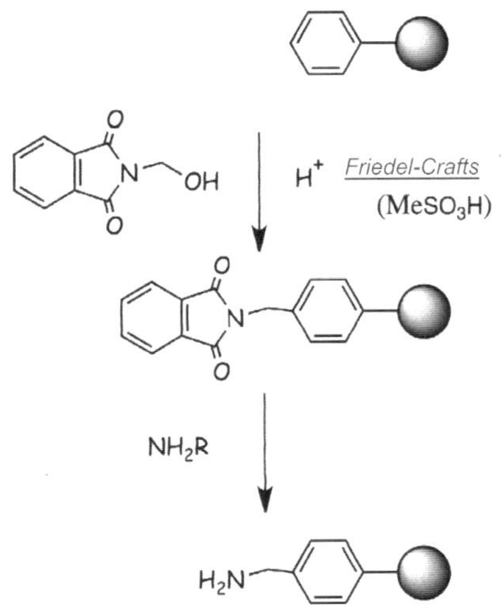

aminomethyl–resin

Paul W. R. Harris et al. ,
Tetrahedron Letters
52 (2011) 6024–6026

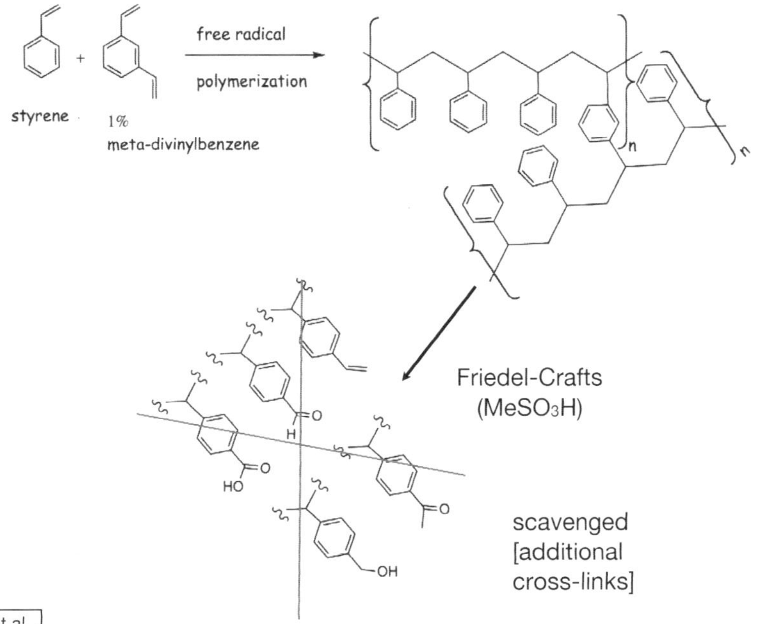

Mitchell A.R., Kent S.B.H., Engelhard M., Merrifield R.B.
J. Organic Chem **43**, 2845-2852 (1978).

Aqueous solvent-compatible resin for SPPS

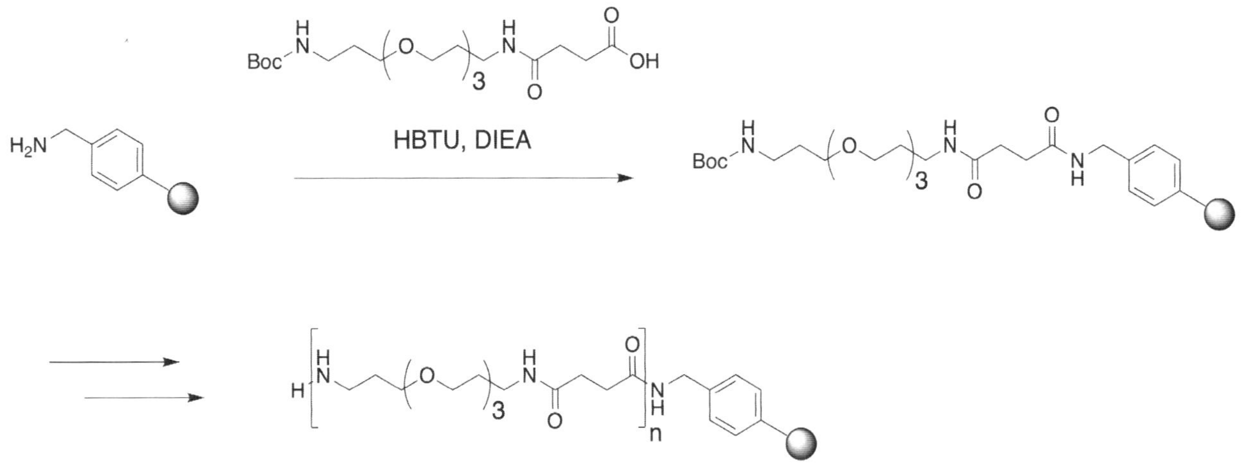

NH_2–(TTD-Succ)$_n$–Resin

Zachary P. Gates, Balamurugan Dhayalan, Stephen B. H. Kent, *Chem.Communications*, 2016,**52**,13979—13982.

Green Solid Phase Peptide Synthesis

Atom economy

- N-carboxyanhydrides
  - minimal side-chain protection
    - Lys(Boc)
    - Ser/Thr?
- high resin loading
  - ~1 millimol/g resin
- maximum ***concentration*** of NCA
- minimum ***excess*** of NCA

Solvent

- water

Resin

- aqueous solvent-compatible

Deprotection/cleavage/purification

- *without* TFA
- efficient purification (PMI)
  - tag-assisted
  - displacement mode prep HPLC

Convergent chemical synthesis

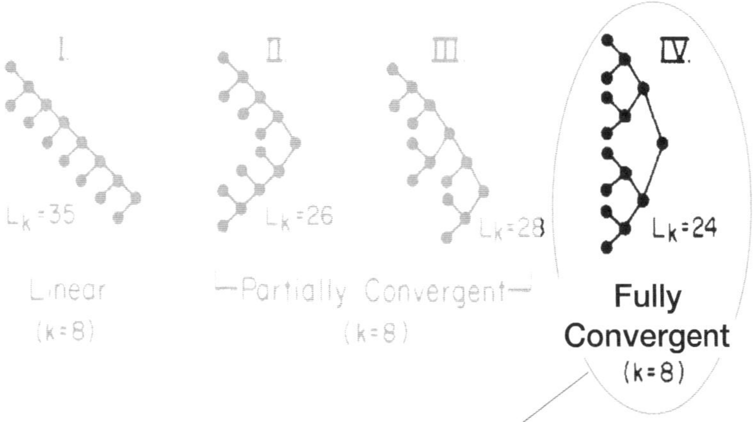

- most efficient use of starting materials
- minimum exposure to reaction conditions
- versatile analogue synthesis
- ***isolation of intermediate products***
- ***enhanced yields & purity***

Systematic Synthesis Design. 6.
Yield Analysis and Convergency

James B. Hendrickson

Journal of the American Chemical Society / 99: 5439 (1977)

Convergent segment ligation *vs.* stepwise synthesis

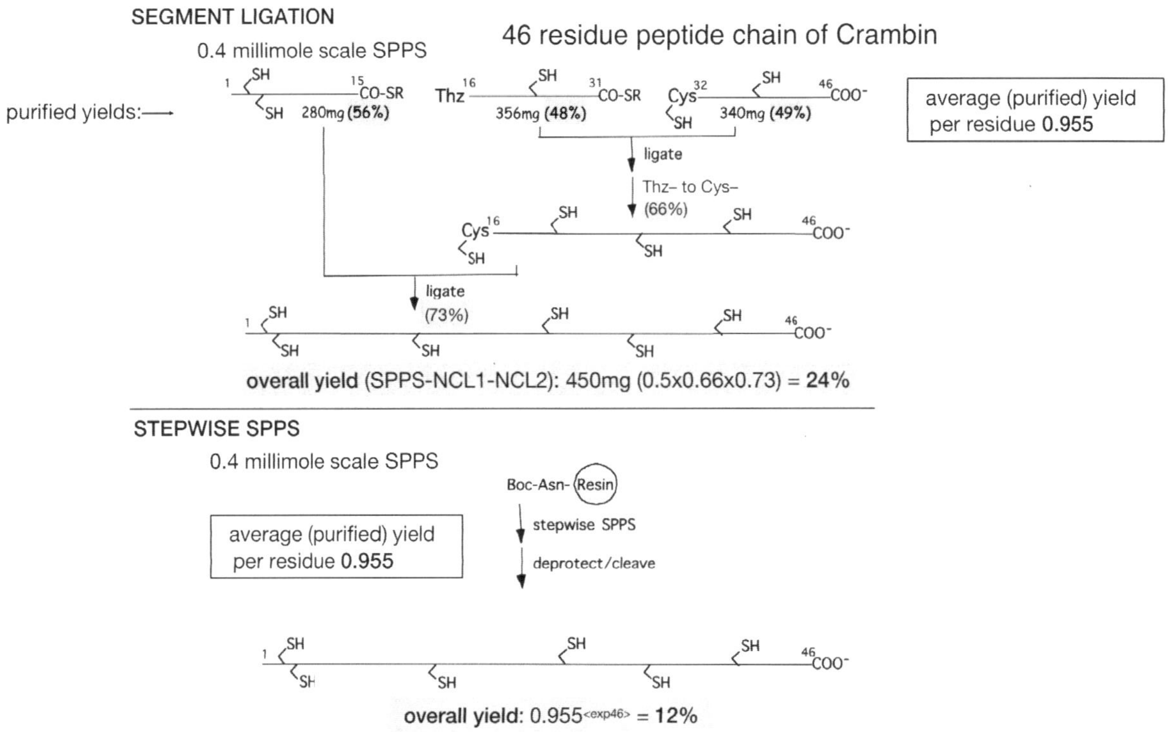

Chemical ligation condensation of unprotected peptides is inherently 'Green'!

* Protecting groups - minimal/none
* Reaction conditions - benign
* Solvent - water (w or w/o 'solubility modifiers')
* Energy - reactions at/near room temperature
* Efficiency - minimal excesses of reactants
* Atom economy - complete/very good
* Products - degradable

*****commercial availability*** of β-SH Xaa's**

– please! –

Green peptide synthesis

stepwise SPPS using N-carboxyanhydrides

convergent segment condensation by NCL

- enhanced yields
- enhanced purity

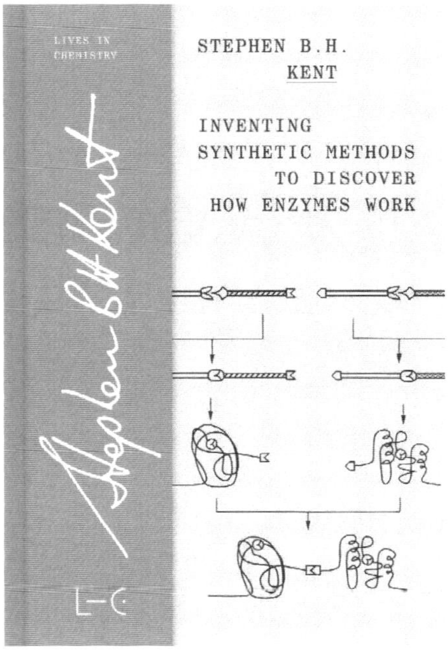
LIVES IN CHEMISTRY
STEPHEN B.H. KENT
INVENTING SYNTHETIC METHODS TO DISCOVER HOW ENZYMES WORK

https://doi.org/10.17952/37EPS.2024.L020

tCAP conjugate: an affinity peptide-based antibody conjugation system to generate highly functional antibody

Yuji Ito[1], Rafique Abdur[1], Shun Masuda[2], Yukie Nohara[2], Shugo Tsuda[2], Taku Yoshiya[2]

[1]Kagoshima University, Kagoshima, Japan, [2]Peptide Institute, Inc., Osaka, Japan

High functional drug with DDS by tissue-specific targeting antibody AccumBody®

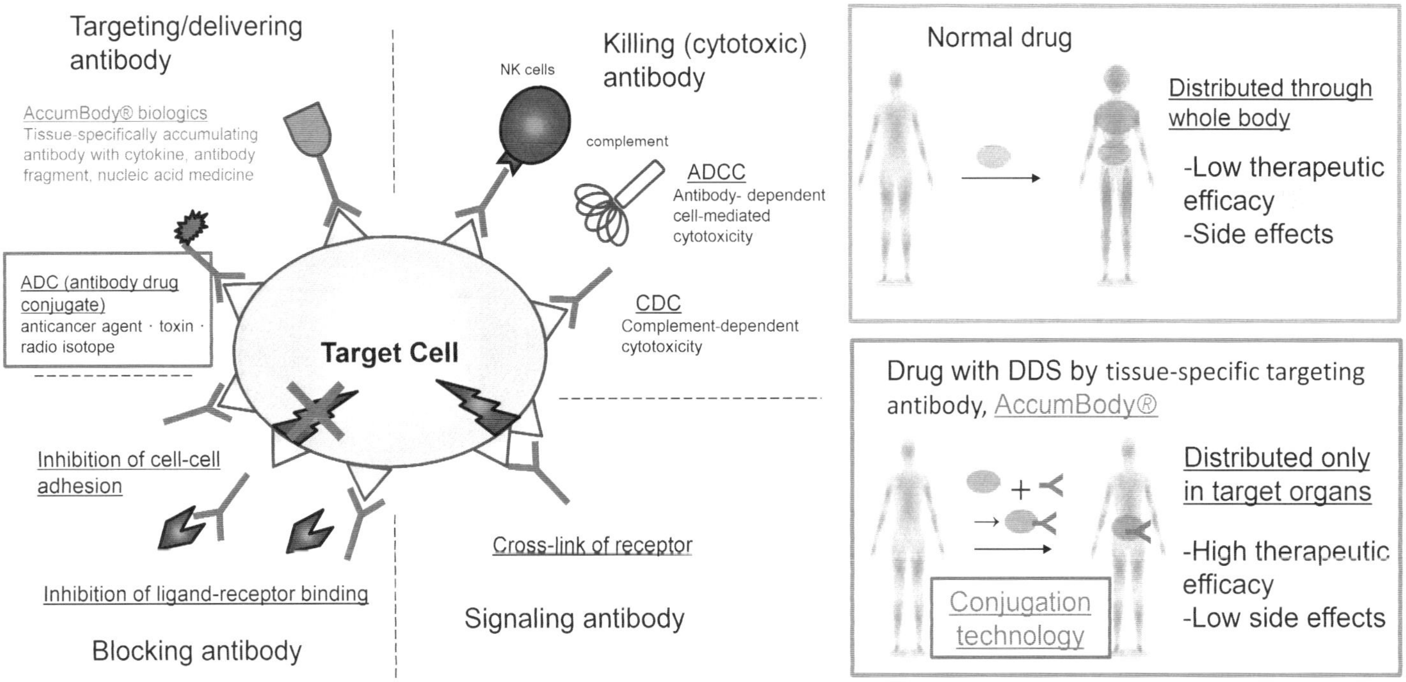

CCAP (Chemical conjugation by affinity peptide)

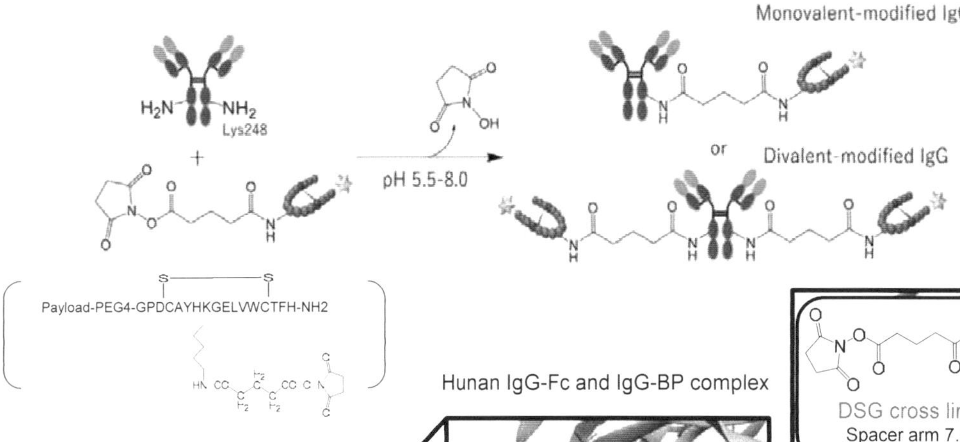

Kishimoto, S. et al. Bioconjug. Chem. 30, 698–702 (2019).

Patent-filed as WO2016186206.

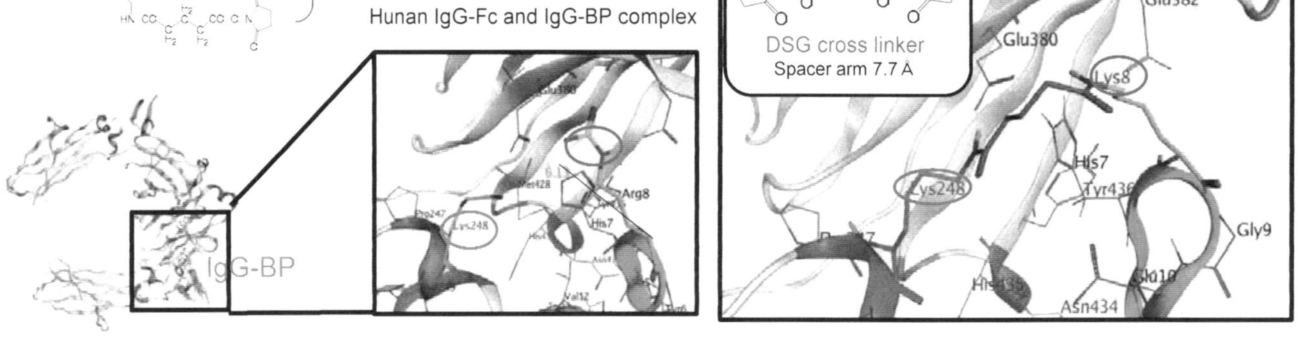

Various ADC prepared by CCAP

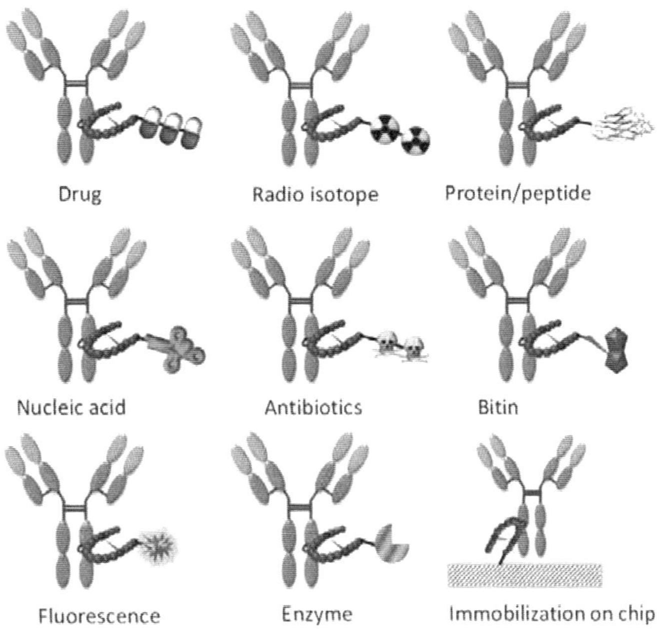

Serum half-life and Fc receptor binding ability of CCAP conjugates

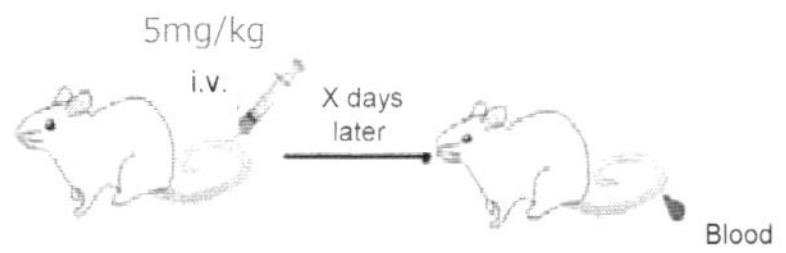

[Dose] 5 mg/kg, I.V.
[Animal] Slc: ICR male, 5 weeks old
[Sampling] 1, 4, 24 h, 2, 3 and 7 d

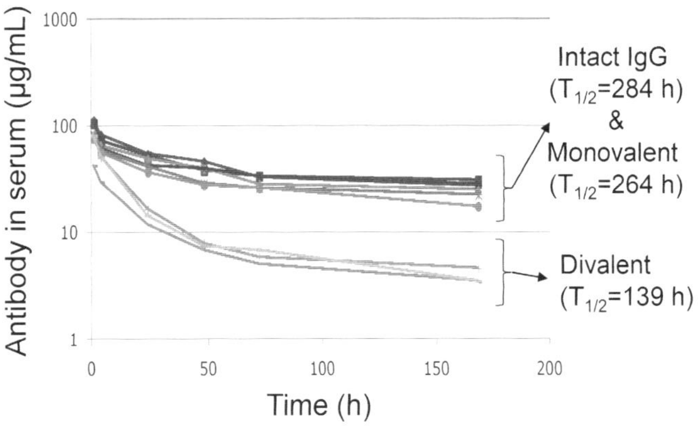

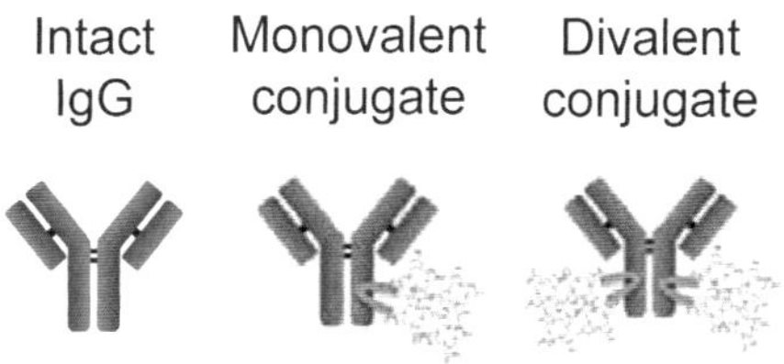

	Trastuzumab (Intact IgG)	Monovalent	Divalent
FcRn	ka=1.2 × 10^{6} (/Ms) kd=4.0 × 10^{-2} (/s) K_D=36 nM	ka=5.6 × 10^{5} (/Ms) kd=3.3 × 10^{-2} (/s) K_D=59 nM	No binding
FcγR I	ka=2.3 × 10^{5} (/Ms) kd=5.7 × 10^{-3} (/s) K_D=24 nM	ka=3.1 × 10^{5} (/Ms) kd=5.4 × 10^{-3} (/s) K_D=17 nM	ka=2.9 × 10^{5} (/Ms) kd=4.8 × 10^{-3} (/s) K_D=16 nM
FcγR Ⅲ A	ka=9.4 × 10^{4} (/Ms) kd=1.5 × 10^{-2} (/s) K_D=160 nM	ka=8.2 × 10^{4} (/Ms) kd=8.4 × 10^{-3} (/s) K_D=100 nM	ka=1.1 × 10^{5} (/Ms) kd=5.0 × 10^{-3} (/s) K_D=45 nM

Next-generation CCAP technology: tCAP

Z34C	:	CH3CO-FNMQCQRRFYEALHDPNLNEEQRNAKIKSIRDDC-NH2
CCAP→ αZ34C	:	DSG-FNMQCQRRFYEALHDPNLNEEQRNARIRSIRDDC-NH2
tCAP→ αZ34C	:	Cargo-MeNbz-Gaba-NMQCQRRFYEALHDPNLNEEQRNARIRSIRDDC-NH2

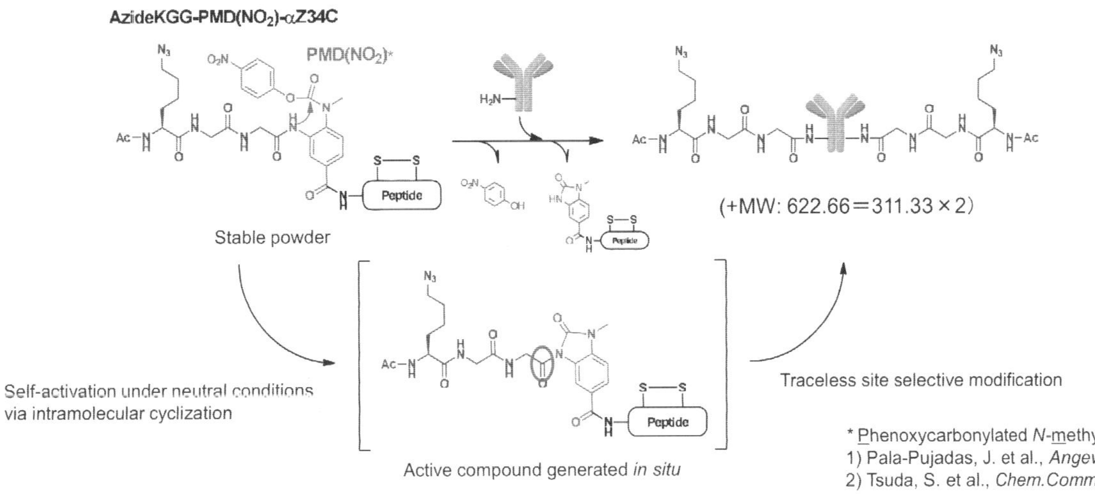

* Phenoxycarbonylated *N*-methyl-diaminobenzene (PMD):
1) Pala-Pujadas, J. et al., *Angew.Chem.Int.Ed.*, **57**,16120 (2018);
2) Tsuda, S. et al., *Chem.Commun.*, **54**, 8861 (2018).

Next-generation CCAP technology: tCAP

Z34C	:	CH3CO-FNMQCQRRFYEALHDPNLNEEQRNAKIKSIRDDC-NH2
CCAP→ αZ34C	:	DSG-FNMQCQRRFYEALHDPNLNEEQRNARIRSIRDDC-NH2
tCAP→ αZ34C	:	Cargo-MeNbz-Gaba-NMQCQRRFYEALHDPNLNEEQRNARIRSIRDDC-NH2

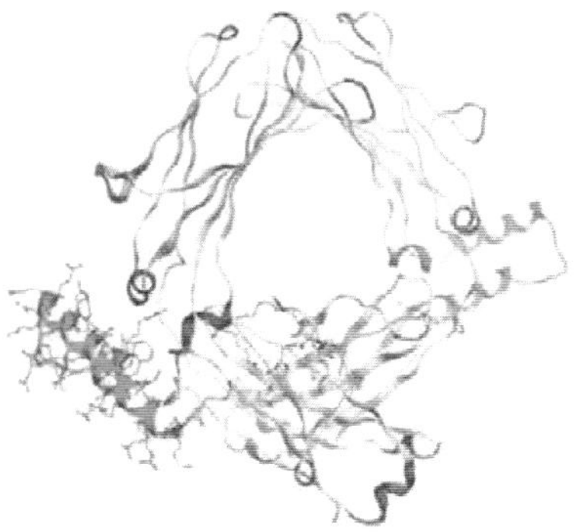

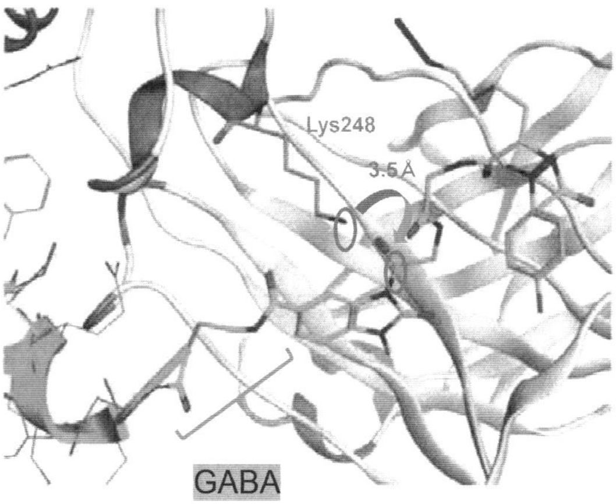

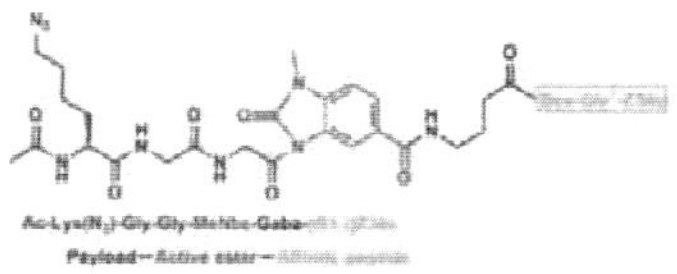

pH dependent reaction of tCAP reagent with human IgG1

Analytical method:

Antibody (15.6 μM) + 5 folds excess reagent
→At different pH and R.T. for 2 h →LC-MS

LC (RP-HPLC):

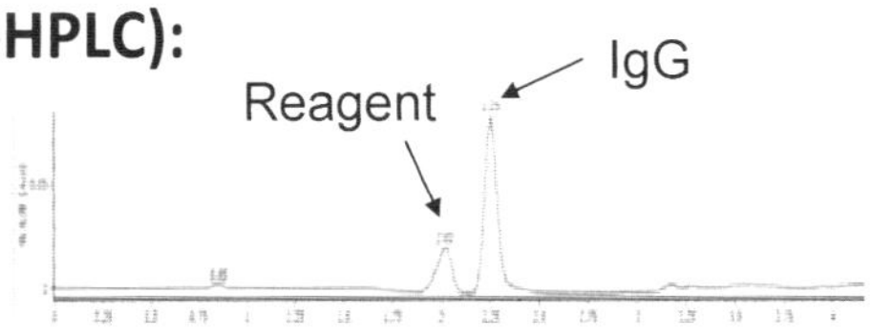

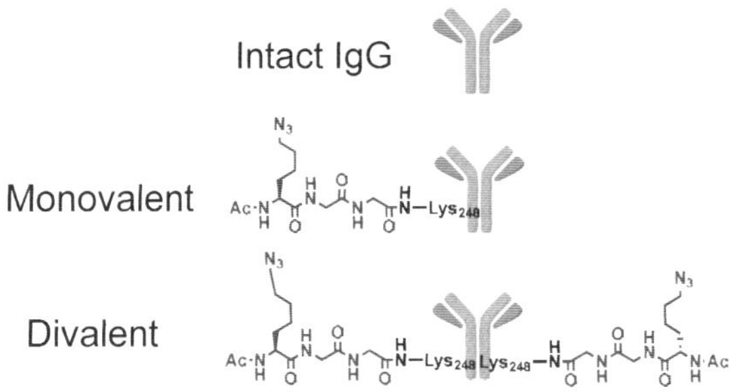

MS of IgG:

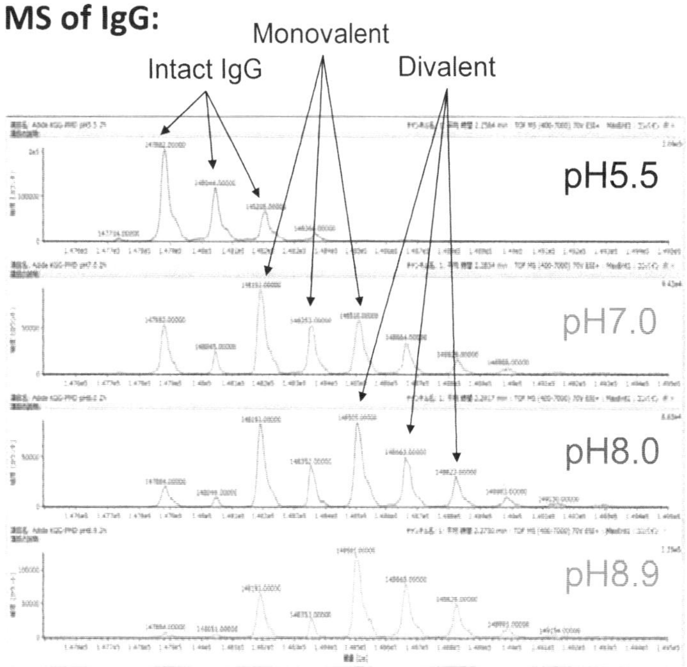

Kinetic binding of tCAP-modified IgG (Trastuzumab) for FcγR and FcRn on BIAcore

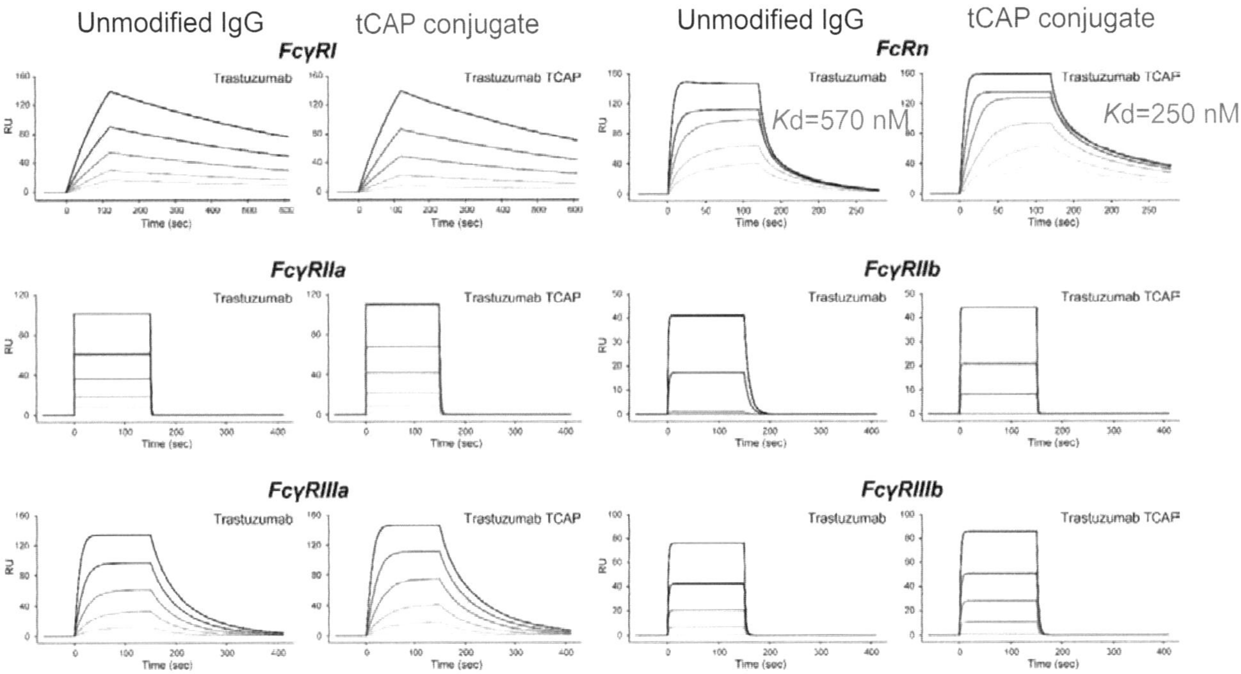

Preparation of Bispecific antibody and AOC by t-CAP

(A) Preparation of conjugate by t-CAP

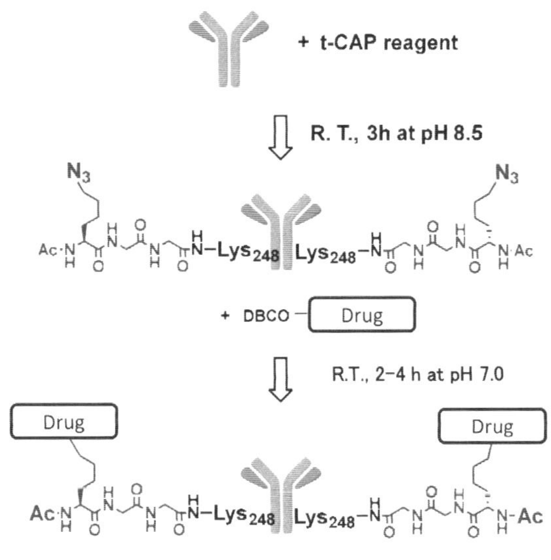

(B) Bispecific antibody
(IgG + VHH)

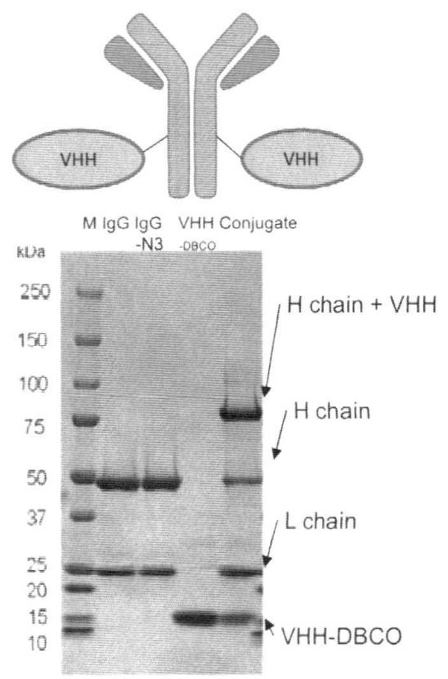

(C) Antibody oligonucleotide
conjugate (AOC)
(VHH-Fc + oligonucleotide)

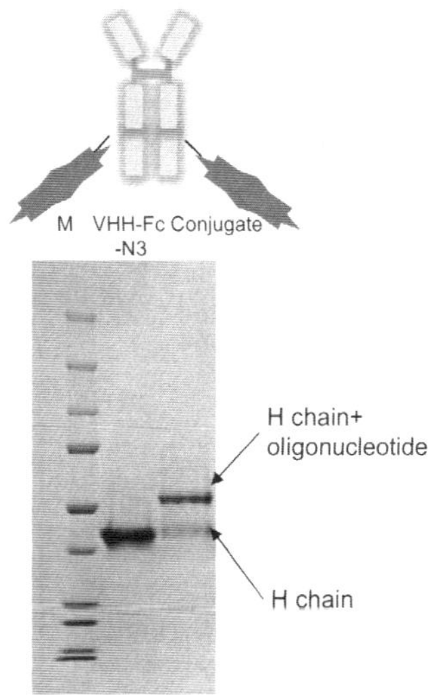

LC-MS analysis of ASO antibody conjugates by tCAP

Attachment of an azide group to 4D5 scFv-Fc by tCAP

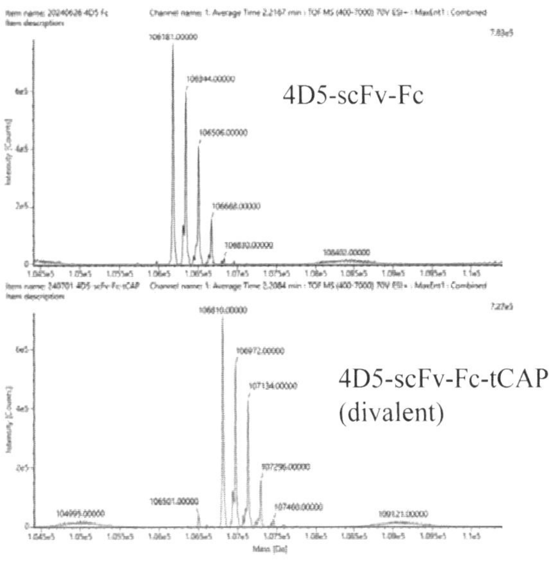

Introduction of oligonucleotide into scFv-Fc antibody by click reaction

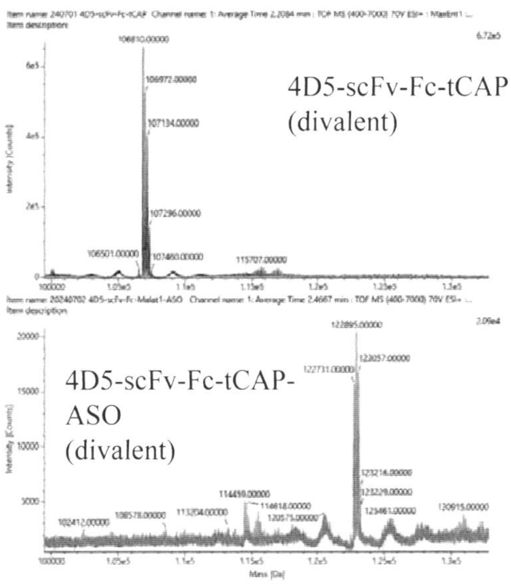

FcRn binding of tCAP conjugates with payloads

FnRn binding evaluation by FnRn- immobilized column*

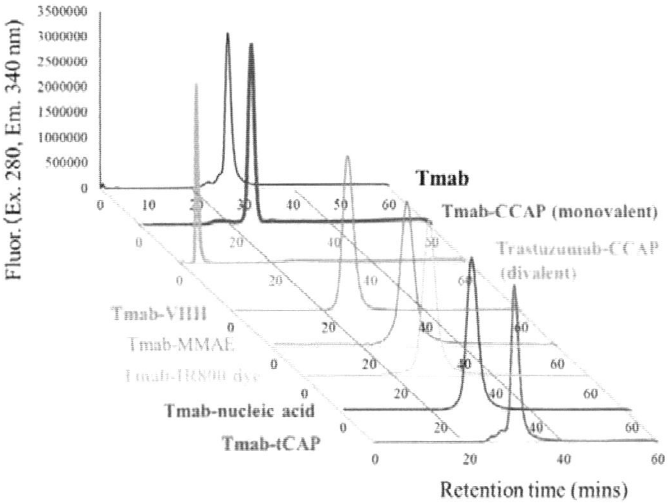

*Bindings with FnRn was evaluated using an FnRn-immobilized column (β version) from Tosoh Corp.

Development of Tissue-Specific Migrating Antibody, AccumBody and Its Application to Next-Generation antibody drug

Drug

Tissue-Specific Migrating Antibody 『AccumBody』

Next-Generation antibody drug

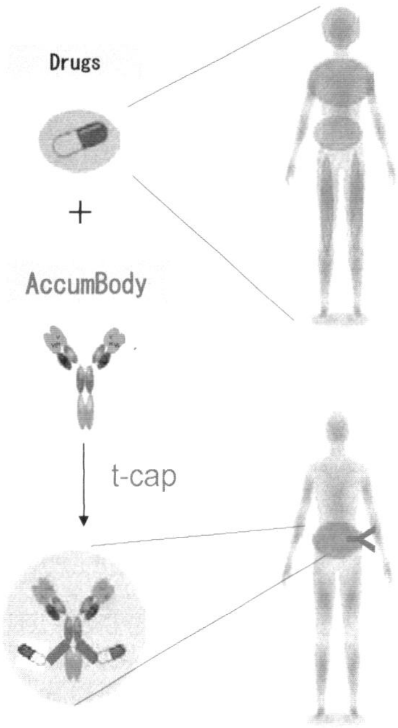

Distributed in whole body

Novel drug candidates for unmet medical needs

C.N. : multiple sclerosis
Muscle : muscular dystrophy
Bowel : IBD

Distributed in target tissue only

- <u>High therapeutic efficacy with small side effects</u>

Collaborators
Professor Y. Kazuki (Tottori Univ.)
Professor K. Tomizuka (Tokyo Phar. Univ.)

Isolation of AcuumBody® by antibody phage library

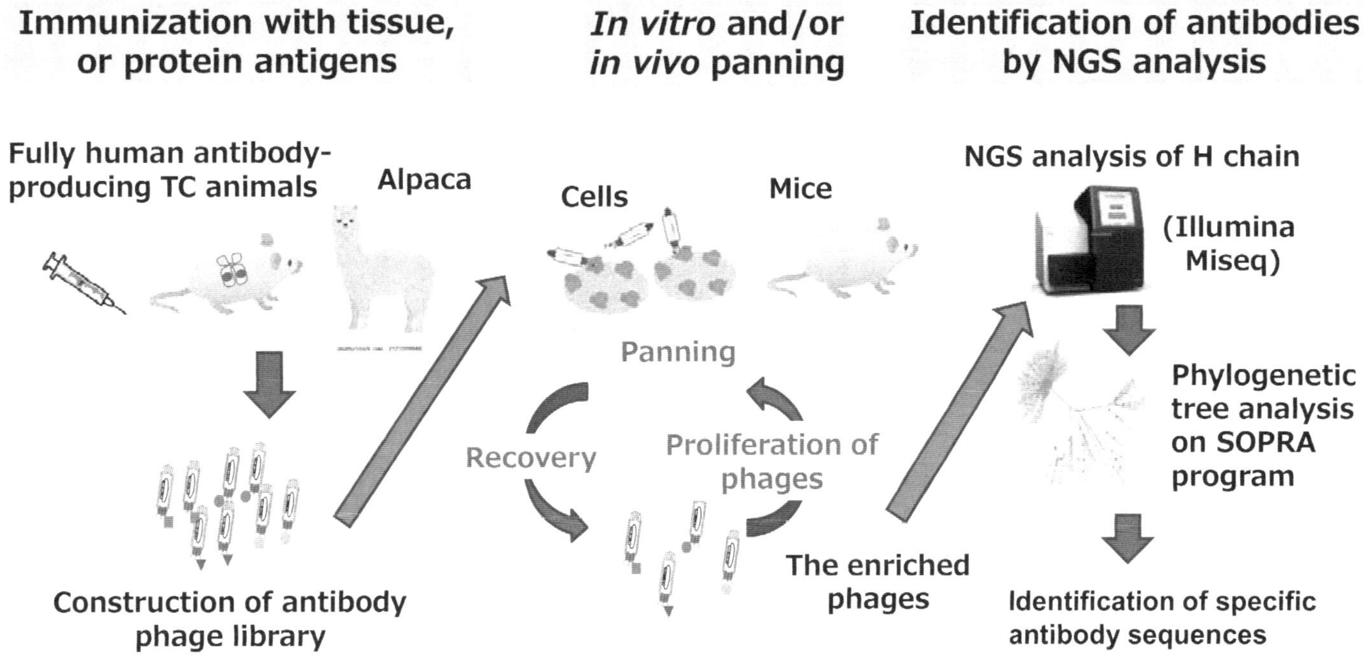

Tissue specific targeting AccumBody®

AcuumBody®-Bowel
(Anti GPA33 antibody)

Cont. Ab

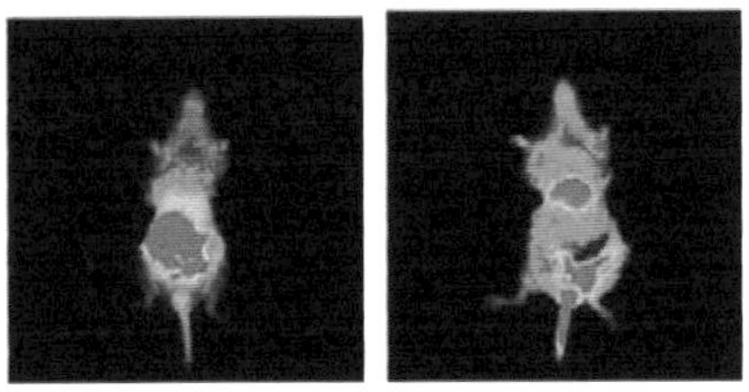

24H

After tail vein administration to Balb/c mice at 5 mg/kg, the mice were perfused and opened at 24 hours for fluorescence imaging.

AccumBody®-Muscle

Cont. Ab

8H

48H

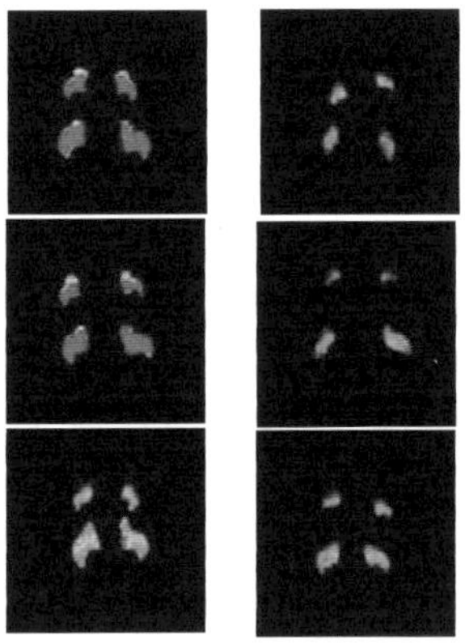

Balb/c mice were perfused at each time after i. v. injection at 5 mg/kg and the muscles of the upper and lower limbs were subjected to fluorescence imaging.

Summary

1. tCAP, a site-specific modification method using antibody Fc-affinity peptides was developed. This method is a useful modification method for the preparation of ADCs, because it does not affect the antigen binding and Fc function including Fc receptor bindings.

2. AccumBodies that target antigens specifically expressed in the brain, bowel and muscle were developed. AccumBodies can deliver the drug to diseased tissues and are useful as DDS tools for highly effective therapeutics.

3. For AccumBody-bowel and -muscle, the technology to promote cellular uptake are necessary to develop. A common issue for the future is the development of cleavable linkers for ADCs.

https://doi.org/10.17952/37EPS.2024.YIF02

Nathan PICOIS

Supervisors : Pr T. Brigaud, Pr G. Chaume et Dr P. Milbeo

***BioCIS** laboratory (UMR CNRS 8076) - Pr T. Brigaud*

Bimodal Use of Chiral α-Trifluoromethylalanine in Aib Foldamers: Study of the Position Impact Towards the Helical Screw-Sense Preference

Young Investigators Mini Symposium flash communications – *29th August 2024*

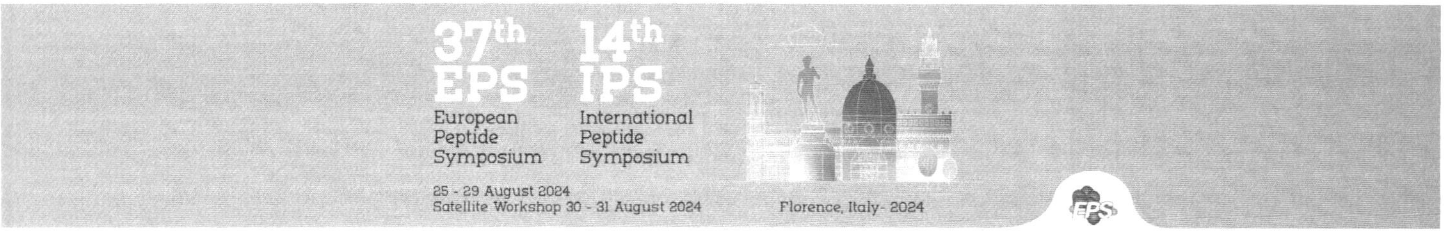

Context and results of the first study

α-Aminoisobutyric acid (Aib) oligomers

- Formation of stable right- (*P*) or left-handed (*M*) 3_{10} helices

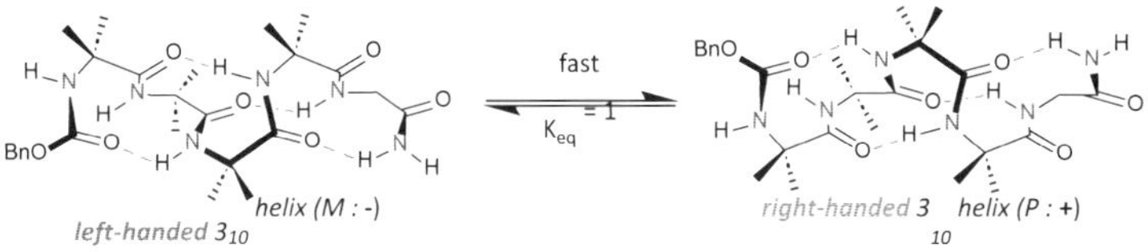

Nature **2022**, *607*, 387-392.

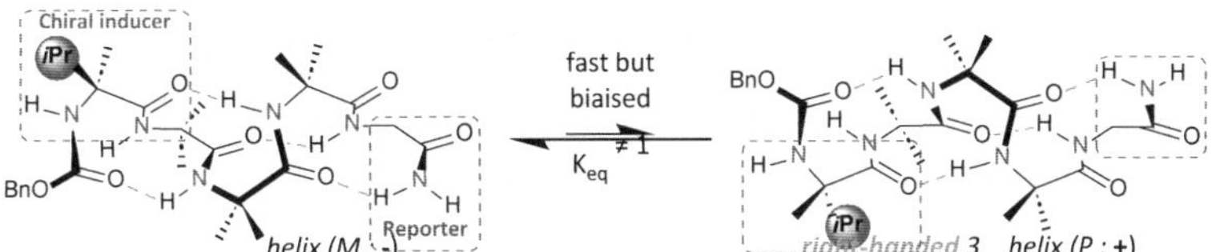

helical excess (h.e.) = **+95%**

J. Org. Chem. **2014**, *79, 10, 4659–4675.*

The chiral inducer promotes a screw-sense preference

X These systems need independently a **chiral inducer** along with an NMR **reporter**

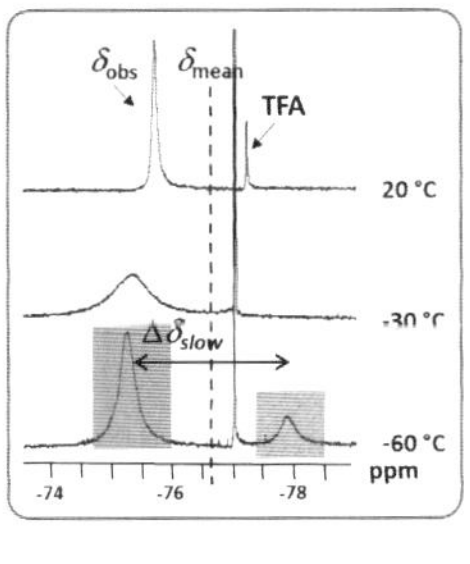

h.e. by ^{19}F **NMR** fits with results obtained through ^{1}H NMR **reporter**

✓ α-TfmAla as a **chiral inducer** and ^{19}F NMR **reporter**

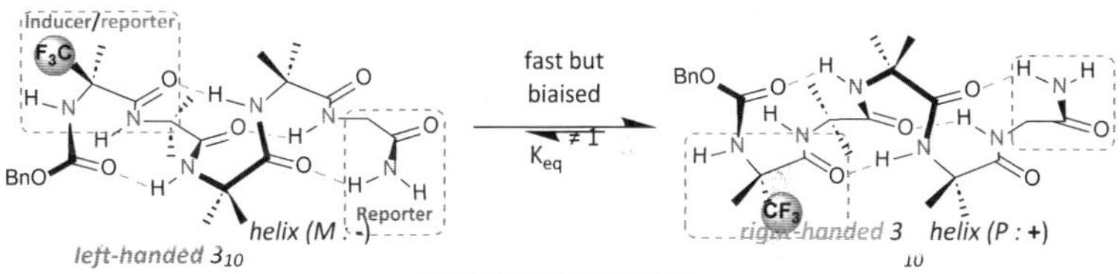

helical excess (h.e.) = **-80/-83%**

Chem. Eur. J. **2022** e202103887.

Bimodal use of α-TfmAla : study of the position impact

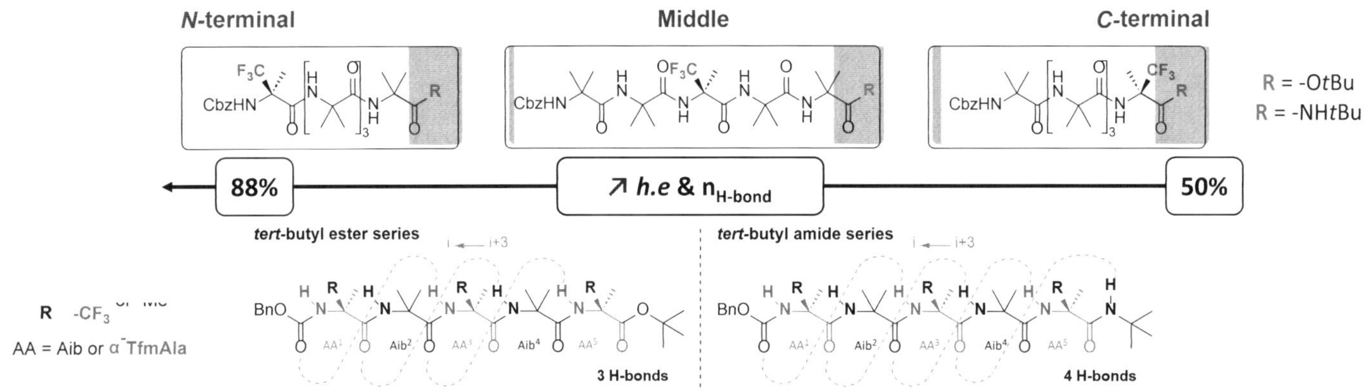

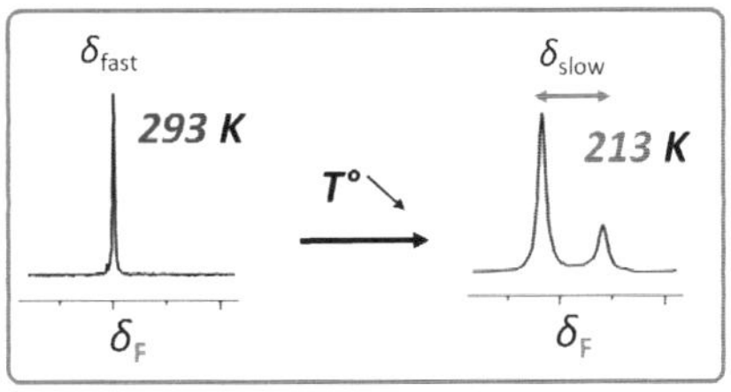

^{19}F NMR allows **helical excess determination at low T°**

CD allows **screw-sense preference determination**

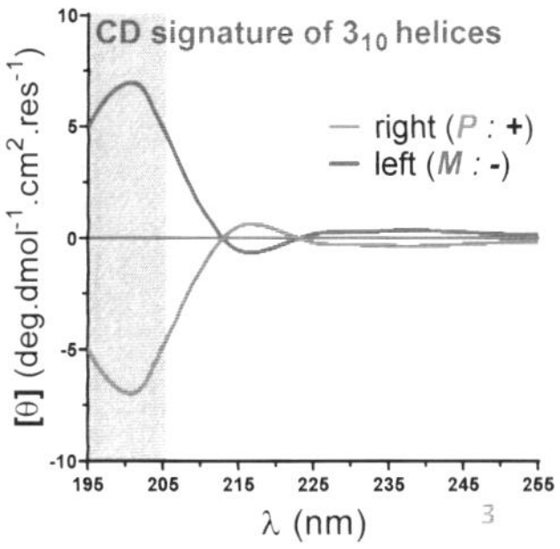

YIMS - 37th European & 14th International Peptide Symposium – Florence, Italy, 29th August 2024

Chem. Eur. J. **2024**, *30*, e202400540.

3

Helical screw-sense inversion : dipole alignment of the –CF_3 group

Chem. Commun. **2014**, *50*, 7949-7952

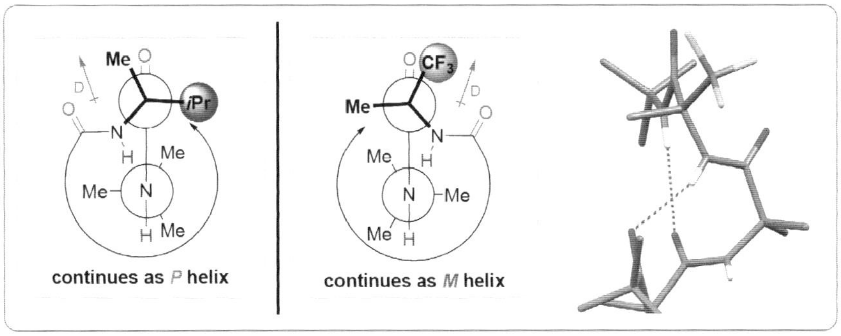

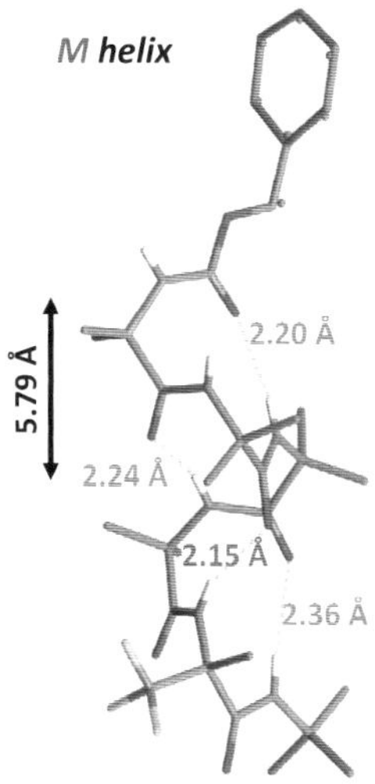

- Stereo-electronic effect of -CF_3 group seems to be the principal contribution
- Dipole alignment observed in the solid state
- ✓ Versatility of **α-TfmAla** residue to induce and determine helical excess in Aib oligomers

Want to learn more ?

Chem. Eur. J. **2024**, *30*, e202400540.

Proceedings of the 37th European Peptide Symposium
Michal Lebl, Editor
European Peptide Society, 2024
http://doi.org/10.17952/37EPS.2024.YIF09

Designed cyclic peptides as inhibitors of amyloid self-assembly and cross-seeding of amyloid-beta peptide and islet amyloid polypeptide

Beatrice Dalla Volta[1], Melanie Roth[1], Yuan Tian[2], Alessia Calzi[1], Jürgen Bernhagen[2], and Aphrodite Kapurniotu[1]

[1]*Division of Peptide Biochemistry, TUM School of Life Sciences, Technical University of Munich (TUM), Freising, D-85354, Germany;* [2]*Division of Vascular Biology, Institute for Stroke and Dementia Research (ISD), Klinikum der Universität München, Ludwig-Maximilian-University (LMU), D-81377, Munich, Germany*

Introduction

Increasing evidence suggests a link between the pathogenesis of Alzheimer's disease (AD) and type 2 diabetes (T2D). In this context, the key amyloid polypeptides of AD and T2D amyloid-β peptide (Aβ (Aβ40(or 42)) and islet amyloid polypeptide (IAPP) have been reported to co-localise in brain or pancreas of AD and T2D patients [1,2]. In addition, IAPP or Aβ fibrils were found to act as cross-seeds of the amyloid self-assembly of the two polypeptides [3,4]. Therefore, molecules that inhibit amyloid self-assembly and cross-seeding of IAPP and Aβ may offer promising leads for anti-amyloid drugs in both diseases.

Our earlier studies revealed high affinity interactions between Aβ and IAPP and were consistent with cross-interactions and amyloid self-assembly being mediated by the same "hot segments" [5,6]. We used the identified IAPP "hot segments" to design peptides as inhibitors of amyloid self-assembly of IAPP, Aβ, or both polypeptides [7,8]. Our more recently developed IAPP-derived amyloid inhibitors were the 17-residue long macrocyclic peptides termed "macrocyclic inhibitory peptides" or MCIPs [8]. The lead MCIP **2e** turned out to be a nanomolar inhibitor of Aβ amyloid self-assembly and in addition it exhibited a good proteolytic stability in human plasma *in vitro* and blood-brain barrier (BBB) crossing in a cell model. However, **2e** was unable to inhibit IAPP amyloid self-assembly. The aim of the work presented here was to design novel cyclic peptides with optimized properties and functions.

Results and Discussion

Systematic MCIP shortening and optimisation yielded a new class of cyclic peptides. Peptides were synthesized by Fmoc-based solid phase peptide synthesis (SPPS) methodology. We used the amyloid specific thioflavin T (ThT) binding assay in combination with transmission electron microscopy (TEM) to study their effects on fibril formation of IAPP or Aβ42. To determine their effects on formation of cytotoxic assemblies, we used the MTT reduction assay in cultured RIN5fm for IAPP-related studies or cultured PC12 cells for Aβ42-related studies. According to the ThT binding assay and TEM, in the presence of the cyclic peptides IAPP fibril formation was fully suppressed. In addition, the cell viability assay showed that formation of cytotoxic IAPP aggregates was fully suppressed as well. Furthermore, the cyclic peptides suppressed Aβ42 amyloid self-assembly while the lead cyclic peptide suppressed IAPP/Aβ42 cross-seeding as well. Thus, the cyclic peptides of our study might be promising candidates for drugs targeting pathogenic amyloid self-assembly in both AD and T2D and studies toward their further development are in progress.

Acknowledgements

This work was supported by Deutsche Forschungsgemeinschaft (DFG) grant SFB1035 B06 to AK (part of SFB1035, project number 201302640) and by DFG under Germany's Excellence Strategy within the framework of the Munich Cluster for Systems Neurology EXC 2145 SyNergy-ID 390857198 to JB. We thank K. Hille and J. Jekel for help with peptide synthesis and purification.

References

1. Jackson, K., et al. *Ann. Neurol.* **74**, 517-26 (2013), https://doi.org/10.1002/ana.23956
2. Martinez-Valbuena, I., et al. *Ann. Neurol.* **86**, 539-551 (2019), https://doi.org/10.1002/ana.25570
3. O'Nuallain, B., et al. *J Biol Chem* **279**, 17490-17499 (2004), https://doi.org/10.1074/jbc.M311300200
4. Moreno-Gonzalez, I., et al. *Mol Psychiatry* **22**, 1327-1334 (2017), https://doi.org/10.1038/mp.2016.230
5. Yan, L.M., et al. *Angew Chem Int Ed Engl* **46**, 1246-1252 (2007), https://doi:10.1002/anie.200604056
6. Andreetto, E., et al. *Angew Chem Int Ed Engl* **49**, 3081-3085 (2010), https://doi.org/10.1002/anie.200904902
7. Andreetto, E., et al. *Angew Chem Int Ed Engl* **54**, 13095-1310 (2015), https://doi.org/10.1002/anie.201504973
8. Spanopoulou, A., et al. *Angew Chem Int Ed Engl* **57**, 14503-14508 (2018), https://doi.org/10.1002/anie.201802979

https://doi.org/10.17952/37EPS.2024.YIF13

Targeting West Nile virus replicases: NS3 and heterodimers inhibitors

PhD student: **Daniele Volpin** [a]

Cristina Peggion [a] (supervisor), Riccardo Rigo [b] (co-supervisor), Mattia Sturlese [b]

[a] Department of Chemical Sciences, University of Padova, 35131, Padova, Italy
[b] Department of Pharmaceutical and Pharmacological Sciences, University of Padova, 35131, Padova, Italy

INTRODUCTION and AIM OF THE PROJECT

The **West Nile virus** (**WNV**) is a flavivirus of the Flaviviridae family. It circulates through a mosquito–bird–mosquito cycle and it is transmitted to humans through Culex mosquitoes. ⟶ Urgency for the **development of new WNV antiviral compounds**.

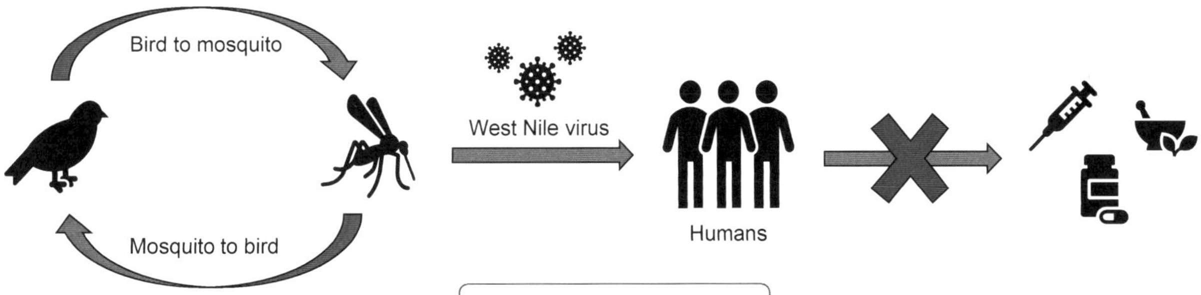

AIM OF THE PROJECT

A **direct approach** targeting the **active site of NS2B-NS3 protease.**

↓

Peptides/peptidomimetics as potential inhibitors.

An **indirect approach** targeting protein-protein interactions essential for NS3 activity.

↓

Peptides to disfavour NS2B-NS3 heterodimer.

NS2B-NS3 ACTIVE SITE INHIBITORS

- **Tri and tetra-peptide inhibitors**

Several peptides and peptidomimetics were synthesized as possible NS2B-NS3 inhibitors of WNV, tested and their IC_{50} value was evaluated.

Peptide name	**Interaction**	**MW (g/mol)**	**Purity**	**IC_{50} (µM)**
Nona-D-Arginine	NON – COVALENT	1422.73	87%	20.70 ± 3.60
PhAc-Lys-Lys-Arg-NH_2		547.71	97%	130.4 ± 8.3
Aun-Lys-Lys-Arg-NH_2		612.86	96%	34.43 ± 3.65
Palm-Lys-Lys-Arg-NH_2		667.99	95%	29.43 ± 3.81
PhAc-Lys-Lys-Arg-(cyclic-dehydro)		514.67	96%	**8.33** ± 0.33
Aun-Lys-Lys-Arg-(cyclic-dehydro)		579.83	98%	**8.34** ± 0.78
Palm-Lys-Lys-Arg-(cyclic-dehydro)		634.95	97%	**10.34** ± 1.71
PhAc-Lys-Lys-Arg-H	COVALENT	532.69	95%	**3.14** ± 0.19
Aun-Lys-Lys-Arg-H		597.85	97%	**7.05** ± 0.97

Tri and tetra-peptide inhibitors

Several peptides and peptidomimetics were synthesized as possible NS2B-NS3 inhibitors of WNV, tested and their IC_{50} value was evaluated.

Peptide name	Interaction	MW (g/mol)	Purity	IC_{50} (μM)
Nona-D-Arg-NH	NON - COVALENT	1422.71	97%	20.70 ± 3.60
PhAc-Lys-Lys-Arg-NH		547.71	97%	[illegible]
Aun-Lys-Lys-Arg-NH		612.86	96%	34.43 ± 3.65
Palm-Lys-Lys-Arg-NH		[illegible]	[illegible]	20.40 ± 3.01
PhAc-Lys-Lys-Arg-(cyclic-dehydro)		514.67	90%	8.33 ± 0.33
Aun-Lys-Lys-Arg-(cyclic-dehydro)		579.83	[illegible]	8.34 ± 0.78
Palm-Lys-Lys-Arg-(cyclic-dehydro)		634.96	97%	10.34 ± 1.71
[illegible]	COVALENT	[illegible]	[illegible]	3.14 ± 0.19
Aun-Lys-Lys-Arg-H		597.85	97%	7.05 ± 0.97

Cyclic peptide inhibitor

Cyclization of the tripeptides to increase the inhibitory efficacy and to disfavor their degradation.

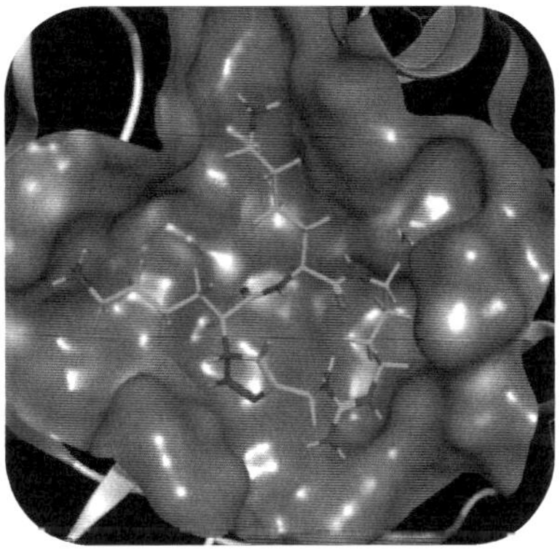

Superimposition of cyclic tripeptide on the active site pocket of NS2B-NS3 protease of WNV.

NS2B-NS3 ACTIVE SITE INHIBITORS

Tri and tetra-peptide inhibitors

Several peptides and peptidomimetics were synthesized as possible NS2B-NS3 inhibitors of WNV, tested and their IC_{50} value was evaluated.

Peptide name	Interaction	MW (g/mol)	Purity	IC_{50} (μM)
Nona-D-Arg-NH2	NON – COVALENT	1422.73	97%	20.70 ± 3.59
PhAc-Lys-Lys-Arg-NH		547.71	97%	[illegible]
Aun-Lys-Lys-Arg-NH		612.86	90%	34.43 ± 3.05
Palm-Lys-Lys-Arg-NH		657.90	95%	20.43 ± 3.81
PhAc-Lys-Lys-Arg-(cyclic-dehydro)		514.67	96%	8.33 ± 0.33
Aun-Lys-Lys-Arg-(cyclic-dehydro)		579.83	98%	8.34 ± 0.78
Palm-Lys-Lys-Arg-(cyclic-dehydro)		634.96	97%	10.34 ± 1.71
PhAc-Lys-Lys-Arg-H	COVALENT	532.69	95%	3.14 ± 0.19
Aun-Lys-Lys-Arg-H		597.86	97%	7.05 ± 0.97

Cyclic peptide inhibitor

Cyclization of the tripeptide to increase the inhibitory efficacy and to disfavor its degradation.

Retro-inverse peptide inhibitors

To increase the inhibitory efficacy and to disfavor its degradation.

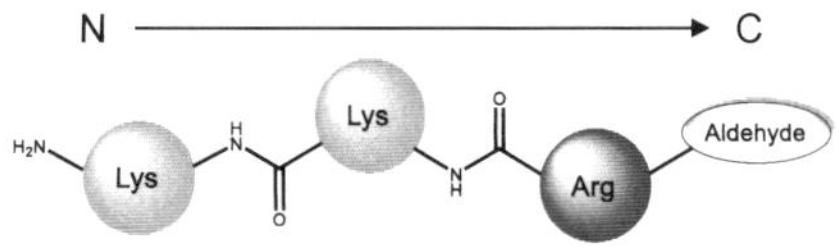

Reference peptide

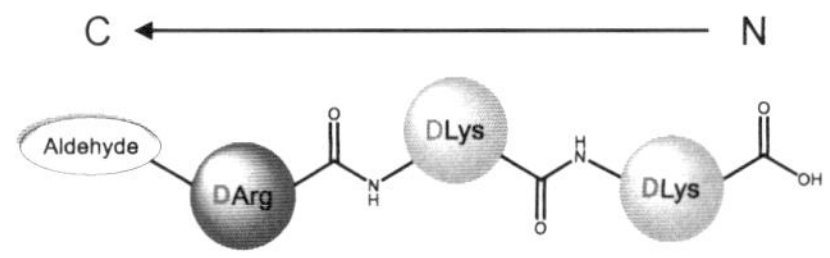

Retro-inverso peptide

We designed **different tripeptides** that mimic the structure of an already known reversible allosteric inhibitor (SID:852843) of the formation of the functional conformation of the NS3 protease of WNV and we studied these **possible allosteric inhibitors by docking**.

- **Active tripeptide allosteric inhibitor**

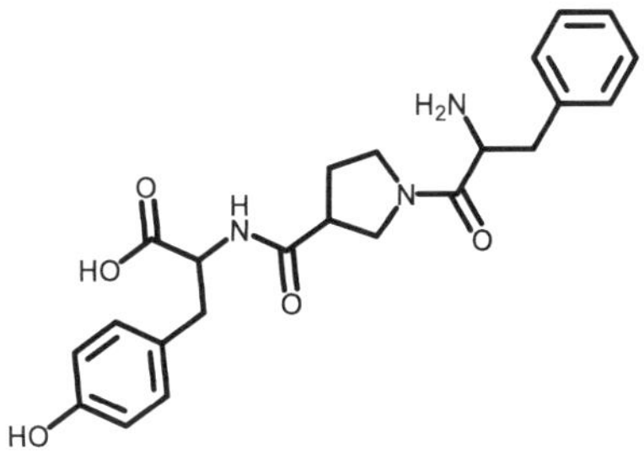

Phe-βPro-Tyr-OH

IC_{50} = **48.30** ± 11.07 μM

NS2B-NS3 ALLOSTERIC SITE INHIBITORS

We designed **different tripeptides** that mimic the structure of an already known reversible allosteric inhibitor (SID:852843) of the formation of the functional conformation of the NS3 protease of WNV and we studied these **possible allosteric inhibitors by docking**.

- Active tripeptide allosteric inhibitor

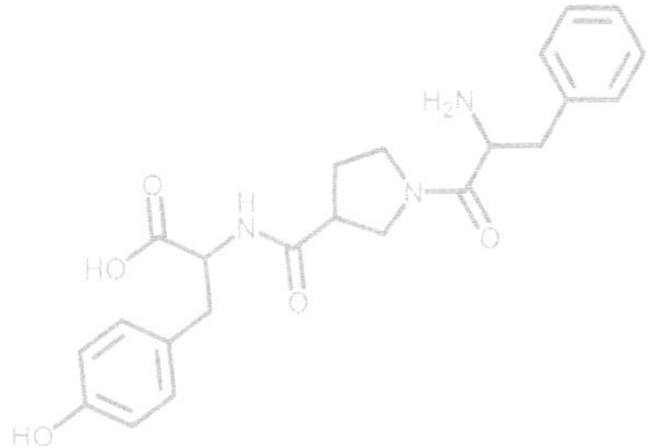

Phe-βPro-Tyr-OH

IC_{50} = **48.30** ± 11.07 μM

- **Protein-peptides interactions by NMR studies**

2D NMR BEST-TROSY spectrum of the Trp-βPro-Trp peptide, which interacts with the NS2B-NS3 protease, as evidenced by Chemical Shift Perturbations.

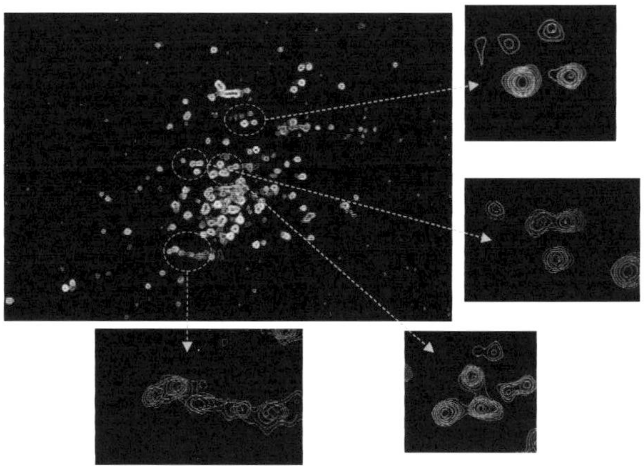

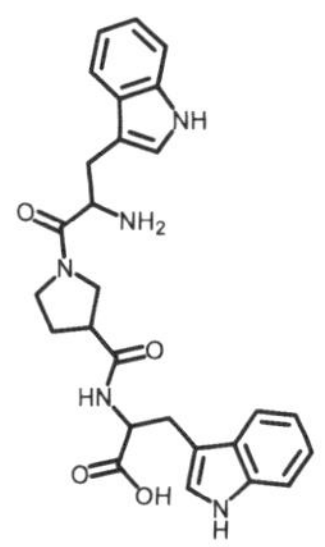

Trp-βPro-Trp-OH

AKNOWLEDGMENTS

Prof. **Cristina Peggion**: supervisor

University of Padova - Department of Pharmaceutical and Pharmacological Sciences

Dr. **Riccardo Rigo**: co-supervisor

Prof. **Mattia Sturlese**

Caterina Zulian
Thomas Stella

University of Padova - Department of Pharmaceutical and Pharmacological Sciences

Prof. Arianna Loregian
Prof. Beatrice Mercorelli

University of Cagliari - Department of life and environmental sciences

Prof. Angela Corona
Prof. Enzo Tramontano

Fondazione INF-ACT

Funded by the European Union
NextGenerationEU

https://doi.org/10.17952/37EPS.2024.YIF20 IIT GUWAHATI

Short Catalytic Peptides for Carbon Dioxide Sequestration

Young Investigators Mini Symposium Flash Communications

37th European Peptide Symposium
14th International Peptide Symposium
25th - 29th August 2024
Florence, Italy

Kalpana Kumari
Doctor of Philosophy

Molecular Informatics & Design Laboratory
Biosciences & Bioengineering
IIT Guwahati, Assam, India
Supervisor: Vibin Ramakrishnan

Motivation and Problem Statement

IIT GUWAHATI

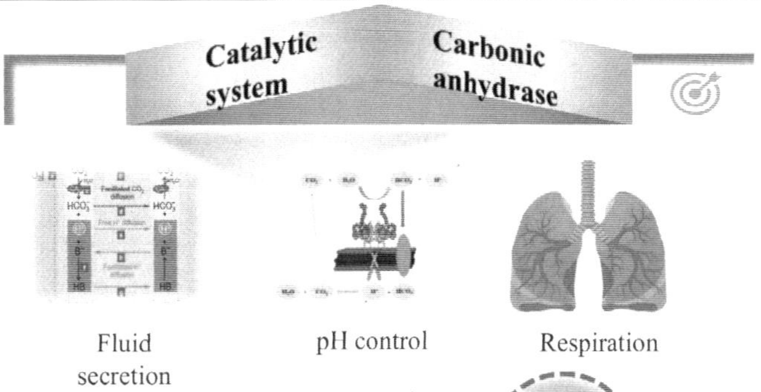

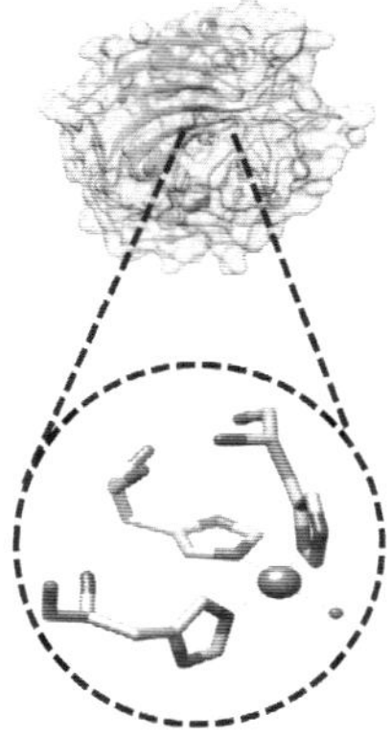

Why peptide catalyst?

1. Self-assembly
2. Inherent biological origin
3. Biocompatible & Biodegradable
4. Low immunogenicity

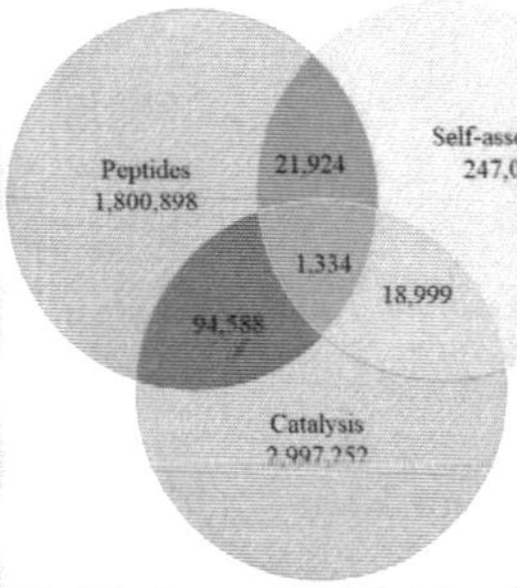

SciFinder, as of 5th March 2018

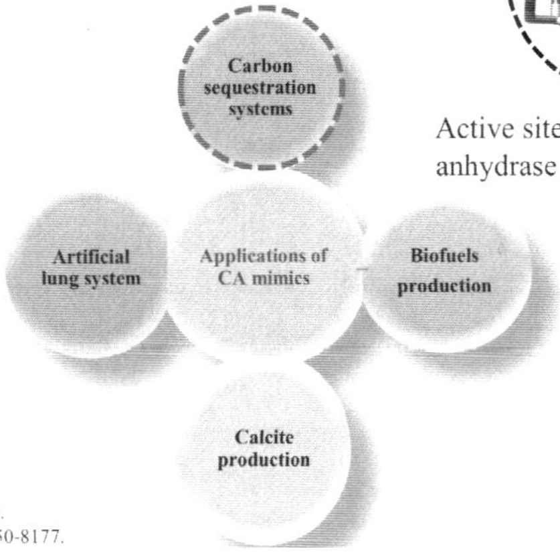

O. Zozulia, M. A. Dolan, and I. V. Korendovych, "Catalytic peptide assemblies," *Chem. Soc. Rev.*, vol. 47, no. 10, pp. 3621–3639, 2018.
Fleming, S., and Ulijn, R. V. (2014) Design of nanostructures based on aromatic peptide amphiphiles, *Chemical Society Reviews 43*, 8150-8177.

Work done on Problem

IIT GUWAHATI

Morphological Characterization of Designed Peptides by Field-Emission Scanning Electron Microscope

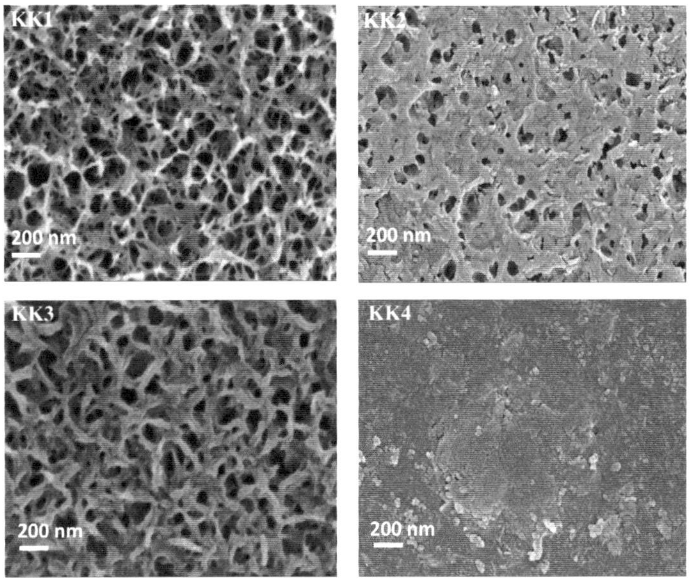

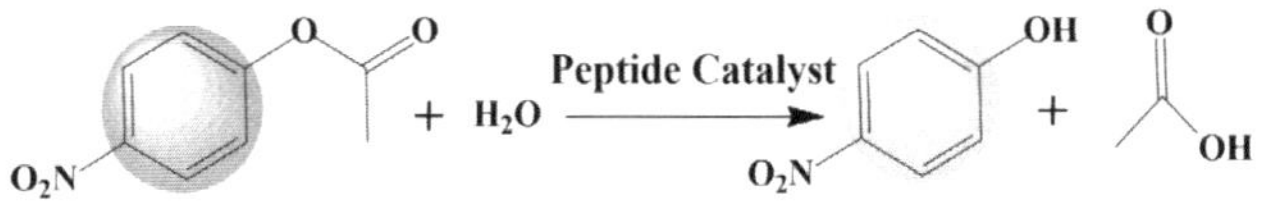

p-Nitrophenyl acetate **p-Nitrophenol**

Kinetic parameters of Peptide Catalysts

Kinetic parameters	KK1	KK2	KK3	KK4
v_{max} (mM/s)	9.39×10^{-4}	0.41×10^{-4}	6.18×10^{-4}	(No activity)
K_M (mM)	0.131	0.329	0.212	-
$K_{cat} = v_{max}$/ Peptide concentration (s^{-1})	93.9×10^{-4}	4.1×10^{-4}	61.8×10^{-4}	-
Catalytic efficiency $= K_{cat}/K_M$ ($M^{-1}s^{-1}$)	71.67	1.246	29.15	-

Patent Filed: R20243014268

Catalytic efficiency: KK1> KK3> KK2> KK4

Work done on Problem

IIT GUWAHATI

Catalytic Activity of Peptide at varying pH and Temperature

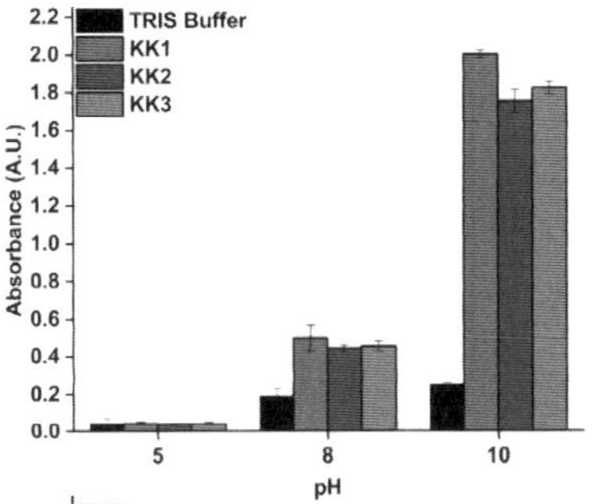

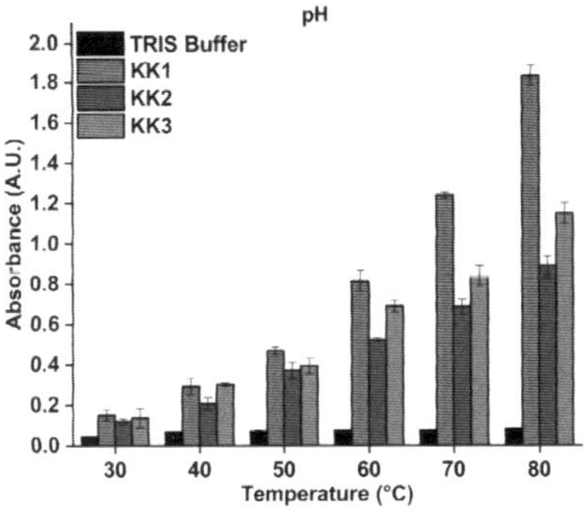

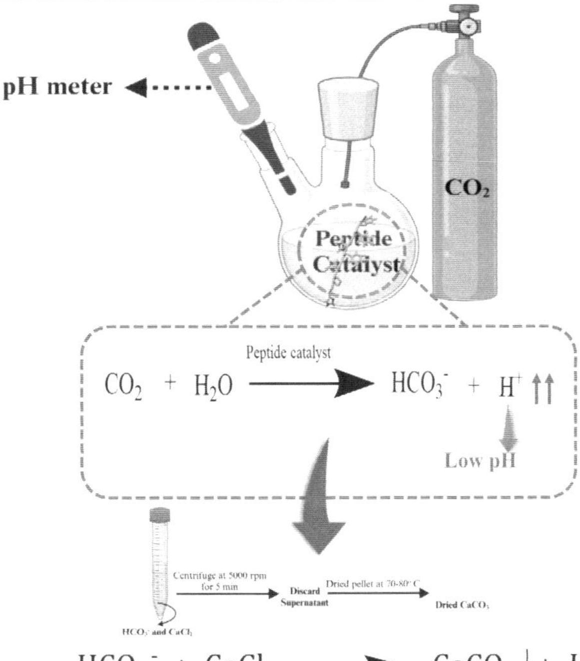

$$HCO_3^- + CaCl_2 \longrightarrow CaCO_3\downarrow + HCl$$

Patent Filed: R20243014268

pH changes during the hydration of CO_2 Followed by Formation of $CaCO_3$ during Sequestration in the Presence of Different Catalysts

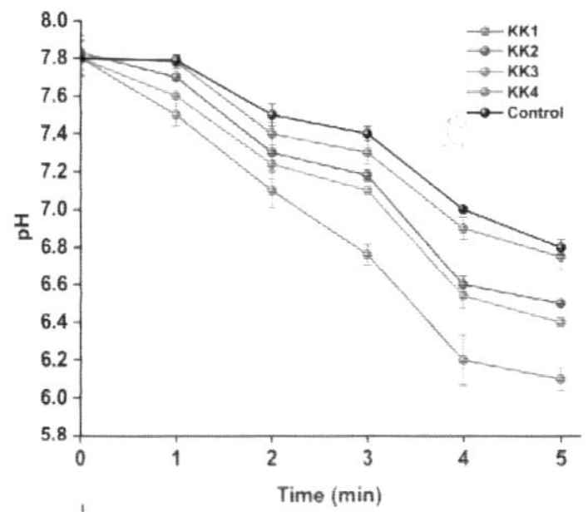

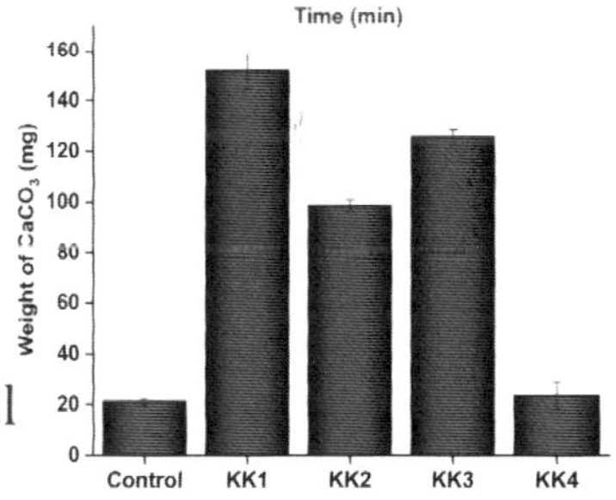

Proceedings of the 37th European Peptide Symposium
Michal Lebl, Editor
European Peptide Society, 2024
http://doi.org/10.17952/37EPS.2024.P1052

N-Aminoimidazol-2-one Peptide Mimic Tools for Deciphering Active Conformations

Yousra Hamdane[1], Pradeep Chauhan[1], Suresh Vutla[1], Darince Truong[1] and William D. Lubell[1]

[1]*Département de Chimie, Université de Montréal, Montréal, Québec H3C3J7, Canada*

Introduction

Peptide mimics serve as tools for studying structure-activity relationships in biomedical applications to improve selectivity and pharmacokinetic properties.[1-3] Among peptide mimics (Figure 1), *N*-aminoimidazolidin-2-one (Aid, **3**)[4] and *N*-aminoimidazol-2-one (Nai, **4**)[5] residues combine properties of the covalent constraint found in α-amino-γ-lactam (Agl, **1**)[6] residues and the stereoelectronic effects of azapeptides (**2**).[7] The Nai residues enable potential for adding ring substituents to mimic amino acid side chains in constrained χ-dihedral angle orientations to improve activity and selectivity.[5]

Results and Discussion

The synthesis of 4-substituted Nai residues was first investigated by a *5-endo-dig* cyclisation of aza-propargylglycine analogs (e.g., **5**, Figure 2).[8] 4,5-Disubstituted Nai analogs were next prepared by diversification of the 5-position using palladium catalysis and Vilsmeier chemistry to install respectively aryl and formyl groups (e.g., **7** and **8**).[9,10] Organic catalysis using azoglycine residues gave later 5-substituted Nai residues which were used to study peptide ligands of the cluster of differentiation-36 receptor (CD36).[11] For example, oxidation of Fmoc-azaGly-Phe-O*t*-Bu (**9**) using *N*-bromosuccinimide gave azopeptide **10**. Subsequent reaction of the azopeptide with aryl acetaldehydes and proline as organocatalyst, followed by acid-mediated dehydration gave 5-aryl Nai dipeptides **11a** and **11b** (Figure 3), which after ester deprotection, were installed into peptides **14a** and **14b** by conventional solid-phase synthesis.[11]

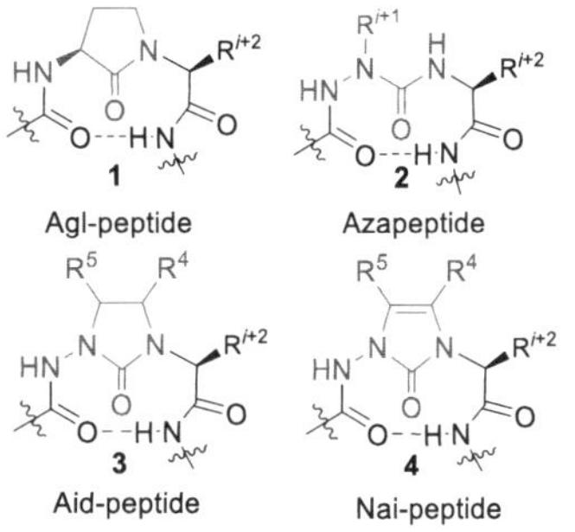

Fig. 1 Agl, aza-, Aid and Nai peptide mimics.

X-ray and computational analyses of model Nai peptides have demonstrated preferences for the *i* + 1 residue of type II' β- and γ-turns as well as potential to mimic side chain function and χ-space orientation contingent upon ring substituent position.[5]

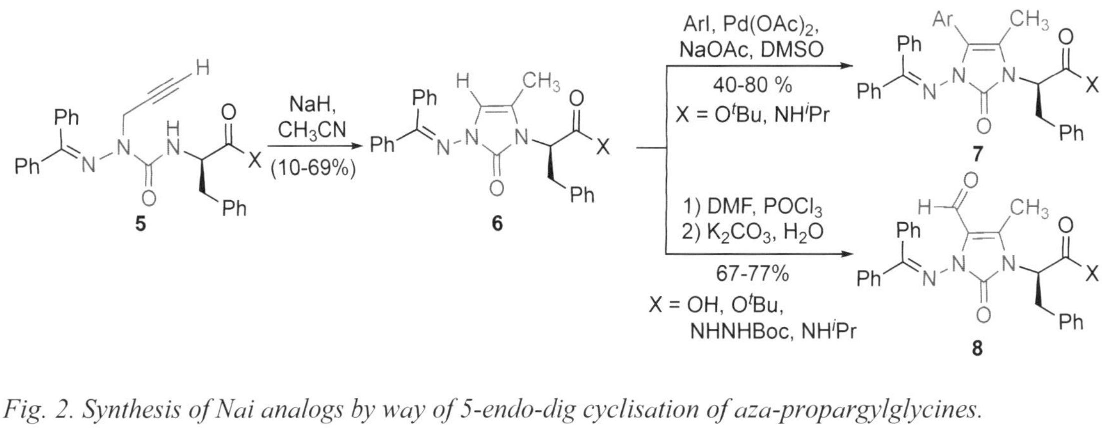

Fig. 2. Synthesis of Nai analogs by way of 5-endo-dig cyclisation of aza-propargylglycines.

Previously, [(5-Ar)Nai[4]]-GHRP-6 analogs **14a** and **14b** exhibited relatively high CD36 binding affinity and reduced NO production by macrophages exposed to a toll-like receptor-2 agonist. The *N*-amino cyclic urea of **14a** and **14b** adopted likely the *i* + 1 residue of a type II' β-turn with aryl substituent in *gauche* χ-space.[11]
With interest to further explore the active conformer, 4,5-disubstituted Nai peptides (e.g., **15**) are under investigation. For example, treatment of azopeptide **10** with 1-phenylpropan-2-one and proline, followed by acid catalyzed dehydration gave (4-Ph,5-Me)Nai **12** in 74% yield from azaGly **9**. Insertion of disubstituted Nai **12** into GHRP-6 analog **15** is now being pursued to detail the active conformation for CD36 modulation towards prototypes for mitigating macrophage-driven inflammation.

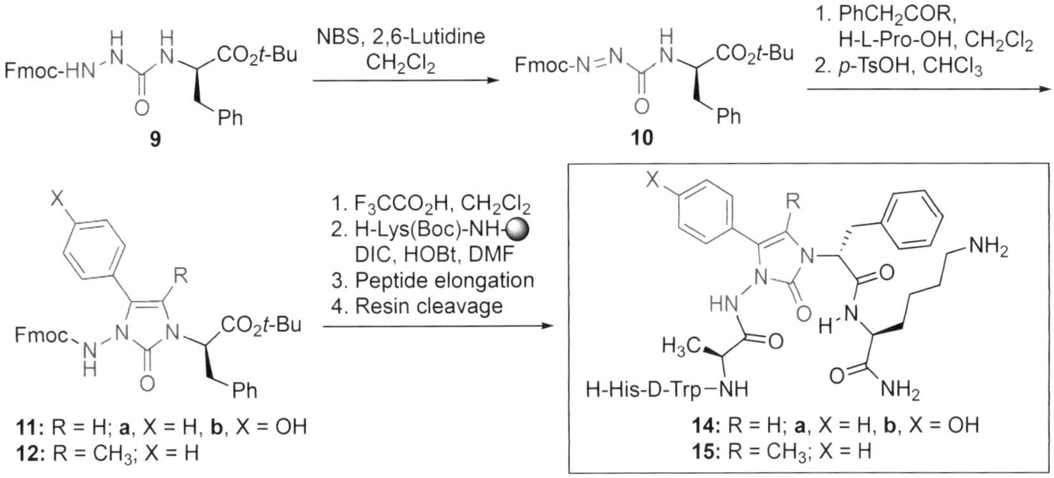

Fig. 3. Organocatalytic approach for Nai peptide synthesis for applications as CD36 modulators.

Acknowledgments

We are grateful for funding from the Natural Sciences and Engineering Research Council of Canada (NSERC, Discovery Research Project RGPIN-2019-04079), the Canadian Institutes of Health Research (CIHR, Project PJT-186296), and the Fonds de Recherche du Québec - Nature et Technologie (FRQNT, CGCC, FRQNT-2020-RS4-265155-CCVC). Assistance is acknowledged from members of the Université de Montréal facilities: Dr. P. Aguiar (NMR spectroscopy) and Dr. A. Fürtös (mass spectrometry).

References

1. Lubell, W. D. *Preface*, In *Peptidomimetics I*; Lubell, W. D., Ed.; Springer International Publishing: Cham, **2017**, v–xiv.
2. Robertson, N. S.; Spring, D. R. *Molecules* **2018**, *23*, 959, https://doi.org/10.3390/molecules23040959
3. Qvit, N.; Rubin, S. J. S.; Urban, T. J.; Mochly-Rosen, D.; Gross, E. R. *Drug Discovery Today* **2017**, *22*, 454, https://doi.org/10.1016/j.drudis.2016.11.003
4. Doan, N.-D.; Hopewell, R.; Lubell, W.D. *Org. Lett.* **2014**, *16*, 2232-5, https://doi.org/10.1021/ol500739k
5. Hamdane, Y.; Poupart, J.; Lubell, D. W. *Synthesis* **2022**, *54*, 1518–1526, DOI: 10.1055/s-0040-1719862
6. Freidinger, R. M. *J. Med. Chem.* **2003**, *46*, 5553-5566, https://doi.org/10.1021/jm030484k
7. Proulx, C.; Sabatino, D.; Hopewell, R.; Spiegel, J.; García Ramos, Y.; Lubell, W. D. *Future Med. Chem.* **2011**, *3*, 1139-1164, https://doi.org/10.4155/fmc.11.74
8. Proulx, C.; Lubell, W. D. *Org. Lett.* **2012**, *14*, 4552-4555, https://doi.org/10.1021/ol302021n
9. Poupart, J.; Doan, N.-D.; Berube, D.; Hamdane, Y.; Medena, C.; Lubell, W. D. *Heterocycles* **2019**, *99*, 279-293, DOI: 10.3987/COM-18-S(F)22
10. Poupart, J.; Hamdane, Y.; Lubell, W. D. *Can. J. Chem.* **2020**, *98*, 278-284, https://doi.org/10.1139/cjc-2019-0479
11. Hamdane, Y.; Chauhan, P. S.; Vutla, S.; Mulumba, M.; Ong, H.; Lubell, W. D. *Org. Lett.* **2021**, *23*, 3491-3495, https://doi.org/10.1021/acs.orglett.1c00936

Proceedings of the 37th European Peptide Symposium
Michal Lebl, Editor
European Peptide Society, 2024 *http://doi.org/10.17952/37EPS.2024.P1054*

Ring Pucker Control in β-Prolines

Nassim Maarouf Mesli and William D. Lubell
Département de Chimie, Université de Montréal, 1375 Ave. Thérèse-Lavoie-Roux, Montréal, Québec, H2V 0B3, Canada

Introduction

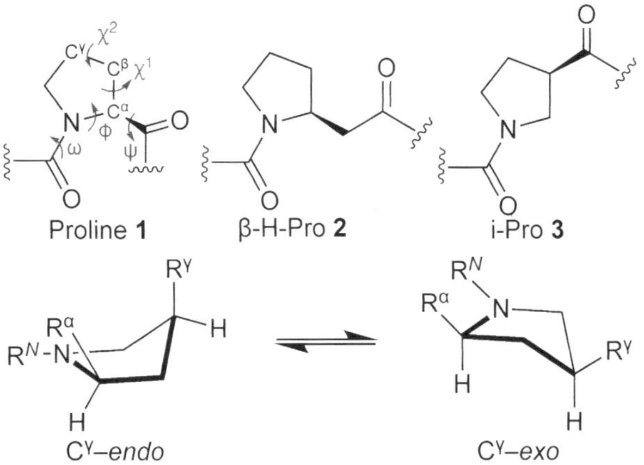

Figure 1. Prolyl and β-prolyl residues and related pyrrolidine ring puckerings.

Pyrrolidine ring puckering influences proline localization in peptide conformations.[1, 2] The dihedral angles χ^1 and χ^2 define respectively the C^γ *endo* ($\chi^1 > 0$, $\chi^2 < 0$) and *exo* ($\chi^1 < 0$, $\chi^2 > 0$) puckers.3-6 In peptides, proline (e.g., **1**) prefers usually the C^γ *endo* ring pucker with the C^α carbonyl in the pseudoaxial orientation (**Figure 1**). Pyrrolidine ring substituents can significantly influence proline puckering and consequently peptide folding.[3-6] For example, (4*R*)-*trans*-hydroxy- and fluoro-prolines favour the *exo* over the *endo* ring pucker in peptides due to stereoelectronic effects.[5] Moreover, *N*-acyl (4*R*)-*trans*- and (4*S*)-*cis-tert*-butyl prolines adopt respectively *endo* and *exo* ring puckers.[6]

The *N*-acetyl analogs of the β-prolines, β-homoproline and isoproline (e.g., **2** and **3**) are predicted to adopt preferred *exo* puckering with pseudoequatorial orientations of the respective C^α and C^β substituents.[7, 8] Exploring the synthesis and conformational preferences of homo-isoproline (Hip) chimeras, we have characterized influences of nitrogen hybridization on ring puckering and peptide topology.

Results and Discussion

The effects of nitrogen hybridization on ring puckering and conformation were examined using di-*N'*-methyl amides of *N*-toluensulfonyl (Ts) and *N*-*p*-nitrophenyl (*p*Np) *cis*-homo-isoprolines *cis*-Ts-Hip(NHMe)-NHMe (**4**) and *cis*-*p*Np-Hip(NHMe)-NHMe (**5**), which were respectively synthesized and characterized by ^{1}H NMR spectroscopy and X-ray crystallography.[9] In sulfonamide **4** the nitrogen is sp^3 hybridized and the *exo* ring pucker is favoured due in part to pseudoequatorial orintations of the C^α and C^γ substituents which minimize steric repulsion (**Figure 2**). Sulfonamide **4** exhibited spectral and crystallographic data for χ^1 and χ^2 dihedral angles in agreement with an *exo* ring pucker as previously reported for 4-substituted prolines.[6] *p*-Nitrophenyl analog **5** is expected to adopt the sp^2 nitrogen hybridization and induce $A^{1,3}$ strain with the C^α substituent which takes a pseudo-axial orientation eclipsing the C^β proton in a flattened C^γ *exo* ring pucker as characterized by X-Ray crystallography (**Figure 2**). The application of the ring puckering effects of homo-isoproline are under further study in peptide mimics.[9]

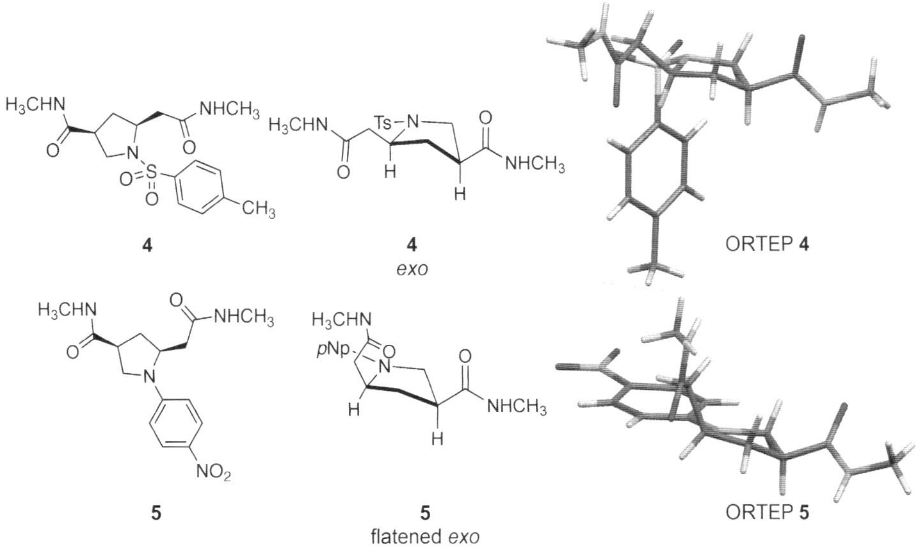

Figure 2. *N*-Substituted homo-isoproline diamides **4** and **5** and favored ring puckerings.

Acknowledgments

We thank the Canadian Institutes of Health Research (CIHR) for project PJT-190158, the Natural Sciences and Engineering Research Council of Canada (NSERC) for funding Discovery Research Project RGPIN-2019-04079, and the Fonds de Recherche du Québec - Nature et Technologie (FRQNT) for support of the Centre in Green Chemistry and Catalysis (CGCC, FRQNT-2020-RS4-265155-CCVC). The authors acknowledge the assistance of members of the Université de Montréal facilities: Dr. P. Aguiar and Dr. C. Malveau (NMR spectroscopy) and Dr. A. Fürtös, Ms. K. Gilbert and Dr. M.-A. Vaudreuil (mass spectrometry), Dr. T. Maris and Dr. D. Chartrand (X-ray). NMM thanks NSERC for an Alexander Graham Bell Fellowship and the EPS for a registration fee waiver.

References

1. Milner-White, E. J.; Bell, L. H.; Maccallum, P. H., Pyrrolidine ring puckering in cis and trans-proline residues in proteins and polypeptides: Different puckers are favoured in certain situations. *J. Mol. Biol.* **1992**, 228 (3), 725-734.
2. DeTar, D. F., Luthra, N. P., Conformations of proline. *J. Am. Chem. Soc.* **1977**, 99 (4), 1232-1244.
3. Berger, G.; Vilchis-Reyes, M.; Hanessian, S., Structural Properties and Stereochemically Distinct Folding Preferences of 4, 5-cis and trans-Methano-l-Proline Oligomers: The Shortest Crystalline PPII-Type Helical Proline-Derived Tetramer. *Angew. Chem. Int. Ed.* **2015**, 54 (45), 13268-13272.
4. Beausoleil, E.; Lubell, W. D., Steric effects on the amide isomer equilibrium of prolyl peptides. Synthesis and conformational analysis of N-acetyl-5-tert-butylproline N '-methylamides. *J. Am. Chem. Soc.* **1996**, 118 (51), 12902-12908.
5. DeRider, M. L.; Wilkens, S. J.; Waddell, M. J.; Bretscher, L. E.; Weinhold, F.; Raines, R. T.; Markley, J. L., Collagen stability: insights from NMR spectroscopic and hybrid density functional computational investigations of the effect of electronegative substituents on prolyl ring conformations. *J. Am. Chem. Soc.* **2002**, 124 (11), 2497-2505.
6. Koskinen, A. M.; Helaja, J.; Kumpulainen, E. T.; Koivisto, J.; Mansikkamäki, H.; Rissanen, K., Locked conformations for proline pyrrolidine ring: synthesis and conformational analysis of cis-and trans-4-tert-butylprolines. *J. Org. Chem.* **2005**, 70 (16), 6447-6453.
7. Góbi, S.; Knapp, K.; Vass, E.; Majer, Z.; Magyarfalvi, G.; Hollósi, M.; Tarczay, G., Is β-homo-proline a pseudo-γ-turn forming element of β-peptides? An IR and VCD spectroscopic study on Ac-β-HPro-NHMe in cryogenic matrices and solutions. *PCCP* **2010**, 12 (41), 13603-13615.
8. Sandvoss, L. M.; Carlson, H. A., Conformational behavior of β-proline oligomers. *J. Am. Chem. Soc.* **2003**, 125 (51), 15855-15862.
9. Mesli Maarouf, N.; Lubell W.D. Unpublished results.

Proceedings of the 37th European Peptide Symposium
Michal Lebl, Editor
European Peptide Society, 2024 *http://doi.org/10.17952/37EPS.2024.P1071*

New unique analogs of anorexigenic cocaine- and amphetamine-regulated transcript peptide

Andrea Pačesová[1], Vilém Charvát[1], Miroslava Blechová[1], Blanka Železná[1], Jaroslav Kuneš[1,2], Lenka Maletínská[1]

1 Institute of Organic Chemistry and Biochemistry of the Czech Academy of Sciences, 160 00 Prague, Czech Republic

2 Institute of Physiology of the Czech Academy of Sciences, 142 00 Prague, Czech Republic

Scientific background: Cocaine- and amphetamine-regulated transcript peptide (CARTp) is an important anorexigenic neuropeptide expressed in the hypothalamus that contains three disulfide bridges important for the biological activity [1]. Its expression is stimulated by leptin, the main regulator of energy homeostasis, which highlights its importance in food intake regulation [2]. CARTp receptor is still unknown, however, Maletínská et al. described the specific binding of CARTp to pheochromocytoma PC12 cells [3]. Subsequent study of Blechová et al. defined the CARTp analog CART(61-102) with only two disulfide bridges, 2-SS-CART(61-102), as the shortest analog with preserved binding affinity and food intake-lowering properties after application to the 3rd brain ventricle [4]. However, natural CARTp is not capable of central action after peripheral administration. Therefore, we designed new lipidized analogs that were tested for their potential anorexigenic properties.

Methods: Lipidized analogs of 2-SS-CART(61-102) were designed and synthesized at the Institute of Organic Chemistry and Biochemistry of the Czech Academy of Sciences in Prague (IOCB CAS in Prague) using Fmoc strategy. The purity and identity of all lipidized analogs were determined by analytical HPLC and using a MALDI-TOF/TOF (Bruker Daltonics, Germany)/Q-TOF micro (Waters) MS technique. The structures are shown in Table 1.

Table 1 Structure of 2-SS-CART(61-102) and its lipidized analogs

Analogue	Sequence
2-SS-CART(61-102)	61 68 74 86 88 94 101 KYGQVPMADAGEQCAVRKGARIGKLADCPRGTSCNSFLLKCL
acyl-2-SS-CART(61-102)	61 68 74 86 88 94 101 Acyl-N-KYGQVPMADAGEQCAVRKGARIGKLADCPRGTSCNSFLLKCL

Acyl: octanoyl (oct), mirystoyl (myr), palmitoyl (palm)

Their binding affinity was tested in PC12 cells, either in naïve cells or those differentiated to neuronal phenotype by nerve growth factor (NGF). The effect on food intake regulation and body weight change was investigate after chronic subcutaneous (SC) application to mice with obesity induced by application of monosodium glutamate (MSG) to newborn mice, day 2-5 after birth, at a dose 4 mg/kg.

Results and discussion: All lipidized analogs showed binding affinity to PC12 cells, naïve or NGF-differentiated, comparable to 2-SS-CART(61-102), as shown in Table 2. palm-2-SS-CART(61-102) showed even higher binding affinity than the natural CART(61-102) with three disulfide bridges [3].

Table 2 Binding affinity of lipidized 2-SS-CART(61-102) analogs

Analog	Naïve PC12 cells	NGF-differentiated PC12 cells
	^{125}I-CART(61–102) K_i [nM]	^{125}I-CART(61–102) K_i [nM]
2-SS-CART(61–102)	261.3 ± 209,80	10.26 ± 9.04
oct-2-SS-CART(61–102)	76.09 ± 17.66	43.9 ± 10.41
myr-2-SS-CART(61–102)	12.1 ± 3.40	14.34 ± 0.33
palm-2-SS-CART(61–102)	1.43 ± 0.41	0.39 ± 0.14

The palm-2-SS-CART(61-102) was subsequently SC administered for 3 weeks to MSG obese mice. The treatment significantly reduced food intake and body weight of the MSG mice. Furthermore, palm-2-SS-CART(61-102) decreased the plasma level of leptin, which is produced by white adipose tissue, and insulin. It is possible that palm-2-SS-CART(61-102) regulates the activity of pancreatic β-cells [5]. In conclusion, we designed new unique CARTp analogs with central effect after peripheral application with strong anorexigenic and antiobesity potential.

Dedication:

This research was supported by the project National Institute for Research of Metabolic and Cardiovascular Diseases (Programme EXCELES, ID Project No. LX22NPO5104) - Funded by the European Union - Next Generation EU, and by RVO: 61388963 and RVO:67985823 of the CAS.

References:

1. Thim, L., *et al.*, *Int J Biochem Cell Biol* **1998,** 30.
2. Kristensen, P., *et al.*, *Nature* **1998,** 393.
3. Maletinska, L., *et al.*, *Eur J Pharmacol* **2007,** 559.
4. Blechova, M., *et al.*, *Peptides* **2013,** 39.
5. Wierup, N., *et al.*, *Peptides* **2022,** 149.

Proceedings of the 37th European Peptide Symposium
Michal Lebl, Editor
European Peptide Society, 2024 http://doi.org/10.17952/37EPS.2024.P1082

Searching for biologically active peptide fragments of hACE2 interacting with the S1 protein of the SARS-CoV-2 virus

Aleksandra Czerchawy[1], Manoj Kumar Muthyala[2], Beata Kolesińska[1]

[1] *Institute of Organic Chemistry, Faculty of Chemistry, Lodz University of Technology, Zeromskiego 116, Lodz, 90-543, Poland* [2] *Gyros Protein Technologies AB, Schillerstraße 20, Teltow, 14513, Germany*

Introduction

Over four years after the COVID-19 pandemic outbreak, we still lack a perfect diagnostic tool for detecting the SARS-CoV-2 virus infection. While the rt-qPCR gained great popularity and became the analytical gold standard, it also has notable drawbacks such as high costs connected with long-term procedures, and the necessity to perform the analysis by well-trained personnel in biosafety level 3 (BSL-3) laboratories. Furthermore, the results are highly dependent on the type and method of sample collection and storage, symptom duration and severity, thus influencing the variable viral load and leading to an increasing number of false-positive results [1]. As an alternative, the rapid test indicating the presence of the virus's spike protein or specific antibodies in the form of LFI tests was used. Nevertheless, they are characterised by lower sensitivity and are typically suitable only for symptomatic individuals, limiting their effectiveness as a broad screening tool.

In response to these challenges, researchers have increasingly focused on developing new types of biomedical tests indicating COVID-19 disease. One promising option is the use of biosensors as small, reliable and portable devices providing real-time results. The receptor layer, typically composed of biological molecules, can ensure high selectivity and specificity with small sample volumes and without the need to isolate the target analyte from the mixture [2]. The covalent immobilization of the biological component offers additionally higher resistance to external factors and allows even for sensor regeneration.

However, the use of protein structures is associated with problematic and time-consuming biosensor development. Their synthesis, isolation and purification are expensive. Additionally, proteins are also prone to enzymatic degradation, thus affecting their stability. A different approach could involve using peptide fragments that maintain the biological activity instead of using the whole protein. Peptides are characterised by higher durability for pH and temperature changes and stability in sterilization and storage conditions. Their smaller size reduces the cost of the synthesis and provides an excellent opportunity for structure modification, enhancing desired properties. What's more, their self-assembling properties can result in forming monolayers with better orientation levels, ensuring also higher density surface, fast binding kinetics and lower risk of aggregation during the interaction, compared with proteins [3]. Fragments mimicking the activity of the receptor would recreate the outer sphere of the protein and be involved in protein-protein or protein-peptide interaction. One of the key receptors, responsible for binding spike protein of the SARS-CoV and SARS-CoV-2 viruses is human angiotensin-converting enzyme type 2 (hACE2).

Result and Discussion

Research conducted in the Institute of Organic Chemistry at the Lodz University of Technology focuses on the development of biosensors indicating the presence of the SARS-CoV-2 virus physiological fluids (e.g. blood, urine, saliva particles, exhaled air) or in the air using hACE2 peptide fragments. The procedure established in our Institute enables the selection of fragments with potential biological activity based on the Dot-blot methodology and involves multiple steps [4].

The initial phase considered the automated SPOT synthesis of overlapping peptide fragments covering the primary structure of the hACE2 on cellulose matrixes with the use of 1,3,5-triazine derivatives (DMT/NMM/TosO$^-$) [5]. Next, the incubation of the received 399-element library with HRP-labelled S1 protein subunit of the SARS-CoV-2 virus and further enzymatic oxidation of 4-chloro-1-naphtol into coloured 4-chloro-1-naphtone allowed the selection of fragments interacting with spike protein (Figure 1).

Fig. 1. *Scans of the cellulose matrix with immobilized decapeptides of hACE2 protein after the incubation with HRP-labelled S1 protein subunit of the SARS-CoV-2 virus. The intensity of each coloured complex represents the strength of the peptide-protein interaction.*

Based on the intensity of the received complexes, measured on a 256-grey scale, 32 structures with different lengths of peptide chains were chosen, including 16 fragments strongly interacting with the spike protein and 16 peptides with moderate affinity (Table 1).

Table 1. Peptide fragments of hACE2 protein selected during the Dot-blot with the use of the HRP-labelled S1 protein subunit of the SARS-CoV-2 virus.

No.	Locant	Peptide fragment	No.	Locant	Peptide fragment
1.	505-526	H-HVSNDYSFIRYYTRTLYQFQFQ-OH	9.	755-776	H-VGIVILIFTGIRDRKKKNKARS-OH
2.	667-692	H-EEDVRVANLKPRISFNFFVTAPKNVS-OH	10.	385-394	H-YAAQPFLLRN-OH
3.	779-788	H-NPYASIDISK-OH	11.	699-718	H-EVEKAIRMSRSRINDAFRLN-OH
4.	545-562	H-SNSTEAGQKLFNMLRLGK-OH	12.	265-282	H-HLLGDMWGRFWTNLYSLT-OH
5.	641-666	H-YLFRSSVAYAMRQYFLKVKNQMILFG-OH	13.	239-256	H-HLHAYVRAKLMNAYPSYI-OH
6.	611-632	H-SPYADQSIKVRISLKSALGDKA-OH	14.	103-122	H-NGSSVLSEDKSKRLNTILNT-OH
7.	733-746	H-PPNQPPVSIWLIVF-OH	15.	467-484	H-EIPKDQWMKKWWEMKREI-OH
8.	587-606	H-YFEPLFTWLKDQNKNSFVGW-OH	16.	123-132	H-MSTIYSTGKV-OH

The further step concerned the resynthesis of chosen fragments using the SPPS method and Fmoc/t-Bu strategy. For some of the fragments, characterized by poor solubility, high hydrophobicity or tendency of aggregation, the process involved automated synthesis with an induction-heating system and "catch-and-release" purification. For this purpose, the coupling of the oxime-based linker to the *N*-terminal amine group of the desired peptide structure was performed. The key benefit of using the linker developed by Zitterbart's team is the rapid, selective immobilization ("catch") of the product to an aldehyde-modified agarose resin. The oxime ligation for the catch step is very robust allowing a wide pH range, the use of organic solvents (ie. dimethylsulfoxide or hexafluoroisopropanol) and chaotropic reagents, thus facilitating the synthesis and purification of "problematic" fragments. After immobilization, any impurities, including by-products, salts and truncated peptides can be washed out. The later traceless cleavage ("release") of the product from solid support is possible through the reduction of an aryl azide to aniline and following acid-catalysed 1,6-elimination of the linker (Figure 2). Moreover, the linker is stable (in its azide form) to acidic environment necessary for peptide cleavage [6].

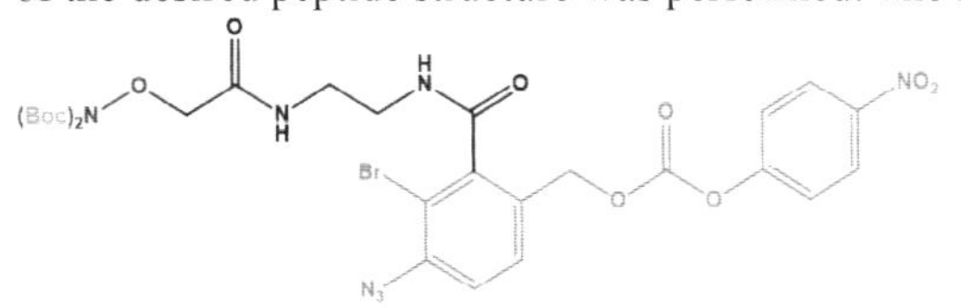

Fig. 2. *The Boc-protected structure of an oxime-based linker. Black colour represents a catch tag (Boc protected), green colours shows a cleavable unit with aryl azide and, brown – leaving group [6].*

The use of the above peptide purification method allowed for obtaining 6 peptide fragments of the hACE2 protein with purities above 90% (Table 2).

Table 2. *Peptides synthesized and purified using an induction heating system and "catch-and-release" method.*

No.	Locant	Peptide fragment	PEC purity
1.	611-623	H-SPYADQSIKVRIS-OH	97,73 %
2.	755-767	H-VGIVILIFTGIRD-OH	95,21 %
3.	385-394	H-YAAQPFLLRN-OH	95,29 %
4.	699-718	H-EVEKAIRMSRSRINDAFRLN-OH	98,85 % *
5.	265-275	H-HLLGDMWGRFW-OH	94,92 % *
6.	111-122	H-DKSKRLNTILNT-OH	96,84 %

* *represents purities including the oxidation of Met residue.*

The last step of the research will include the confirmation of peptides' biological activity towards the SARS-CoV-2 virus particles in MST analysis.

Conclusions

The rapid advancement of natural and technical sciences has led to the development of innovative analytical tools that may find their application in medical diagnostics. Despite the limitations of current diagnostic methods for the SARS-CoV-2 virus identification, promising alternatives are being developed, including biodetecting devices utilizing hACE2 as the receptor structure. The procedure presented in this work shows an alternative approach and may facilitate the process of searching for biologically active peptides, as well as their subsequent synthesis and purification for biomedical application. However, extensive testing would be required to validate the utility and accuracy of such prepared devices.

Acknowledgements

The research was founded by "Young Scientist Fund" no. W-3D/FMN/16G/2023.
I would like to acknowledge the support of Gyros Protein Technologies, in particular Dr. Manoj Muthyala, during the internship at the PECS laboratory in Teltow, Germany, for assisting with the synthesis and PurePep® EasyClean purification of several peptides mentioned in this work. PurePep® EasyClean technology described here can be purchased from Gyros Protein Technologies.

References

[1] Pan, Y., et al. *Clin. Chem.* **66(6)**, 794-801 (2020), https://doi.org/10.1093/clinchem/hvaa091
[2] Barbosa, A.J.M., Oliveira, A.R., Roque, A.C.A. *Trends Biotechnol.* **36 (12)**, 1244-1258 (2018), https://doi.org/10.1016/j.tibtech.2018.07.004
[3] Zhu, Q., Zhou, X. *J. Hazard. Mater.* **425**, 127923 (2022), https://doi.org/10.1016/j.jhazmat.2021.127923
[4] Kolesińska, B. et al. *Engineering of Biomaterials* **148**, 41 (2018).
[5] Frączyk, J., Walczak, M., Kamiński, Z.J. *J. Pept. Sci.* **24**, e3063 (2018), https://doi.org/10.1002/psc.3063
[6] Zitterbart, R. et al. *Chem. Sci.* **12 (7)**, 2389–2396 (2021), https://doi.org/10.1039/D0SC06285E

Proceedings of the 37th European Peptide Symposium
Michal Lebl, Editor
European Peptide Society, 2024
http://doi.org/10.17952/37EPS.2024.P1116

Fragments of fibroblast growth factors useful in regenerative medicine components

Katarzyna Czerczak-Kwiatkowska[a], Adrianna Wierzbicka[b], Dorota Bociaga[b], Justyna Fraczyk[a], Beata Kolesinska[a]
[a]*Institute of Organic Chemistry, Łódź University of Technology, Żeromskiego 116, 90-924 Lodz, Poland*
[b]*Institute of Materials Science and Engineering, Lodz University of Technology, Stefanowskiego. 1/15, 90-924 Lodz, Poland*

Introduction

Tissue engineering and/or regenerative medicine is a field of life science using different methods to create a new tissues/organs and/or to promote the regeneration of damaged or diseased tissues/organs. Materials used as scaffolds must ensure appropriate micro-environment of the cells, adhesion, proliferation, differentiation and finally the reproduction of the parent tissue [1,2]. Concentrated Growth Factors (CGFs) are the newest and safest therapy in regenerative medicine. CGF is a new-generation autologous platelet concentrate containing equivalent or higher levels of TGF-B1, PDGF-BB, and VEGF than platelet-rich plasma (PRP) concentrates. CGFs has been shown to stimulate cell proliferation [3,4], allowing for use in regenerative medicine. [5]. CGF is easily obtained by centrifuging intravenous blood of the patient and does not require any synthetics or biomaterials to make a gel; hence, it is free from the risk of cross-contamination [6]. The fibroblast growth factor (FGF) family represents a broad group of signaling proteins with structural similarities and related actions. FGFs are involved in a number of processes taking place in the cells; they are involved in wound regeneration, angiogenesis or regulation of metabolism, among others. The FGF family consists of 22 proteins. Apart from FGF15 and FGF19, the remaining FGFs have analogous mouse and human structures (homology about 90%). FGFs have very similar biological effects and a similar mechanism of gene expression Fibroblast growth factor 1 (FGF1) is also known as acidic fibroblast growth factor (aFGF). It consists of 155 amino acid residues, the main chain includes residues 16-155. FGF1 occurs as a monomer or homodimer. FGF1 binds to all receptors FGFR (FGFR1-FGFR4). In addition to homodimers, FGF1 forms a complex with fibroblast growth factor binding protein 1 (FGFBP1) and FGF2. Fibroblast growth factor 2 is also known as basic fibroblast growth factor (bFGF). The FGF2 chain consists of 288 amino acid residues, residues 1-142 constitute the propeptide. The main chain includes amino acids 143-288. It occurs as a monomer or homodimer. It binds to FGFR1, FGFR2, FGFR3 and FGFR4 receptors. FGF2 levels are increased in acute wounds, supporting the formation of granulation tissue, re-epithelialization and reconstruction of the damaged site [7]. It also regulates the synthesis and function of ECM components and increases the migration of keratinocytes and fibroblasts [8]. In the case of chronic wounds, the level of FGF2 is reduced [9]. Recombinant FGF2 applied directly to the wound shows promising effects in the treatment of pressure ulcers [10]. Promising effects have also been shown in the treatment of diabetic foot [11]. Currently, derivatives of fibroblast growth factors are available for the treatment of ulcers and burns. FGF proteins are recommended by the American Wound Healing Society and the European Wound Management Association for the treatment of refractory ulcers. Studies on the

 http://doi.org/10.17952/37EPS.2024.P1116

family of these proteins show that FGF2, FGF7 and FGF10 are key in the regeneration of skin wounds.

The aim of the study was to find a set of short fragments of native FGFs involved in repairing damaged tissues. These sets should reproduce the outer sphere of proteins and have biological properties similar to native proteins due to their ability to interact with receptors. We focused our attention on the search for FGF1 and FGF2 fragments.

Results and discussion

Using the SPOT method, libraries of FGF 1 and FGF2 fragments immobilized on cellulose were obtained [12]. Based on the Dot-blot test, fragments of both proteins were selected forming an outer sphere of proteins capable of interacting with the corresponding receptor.

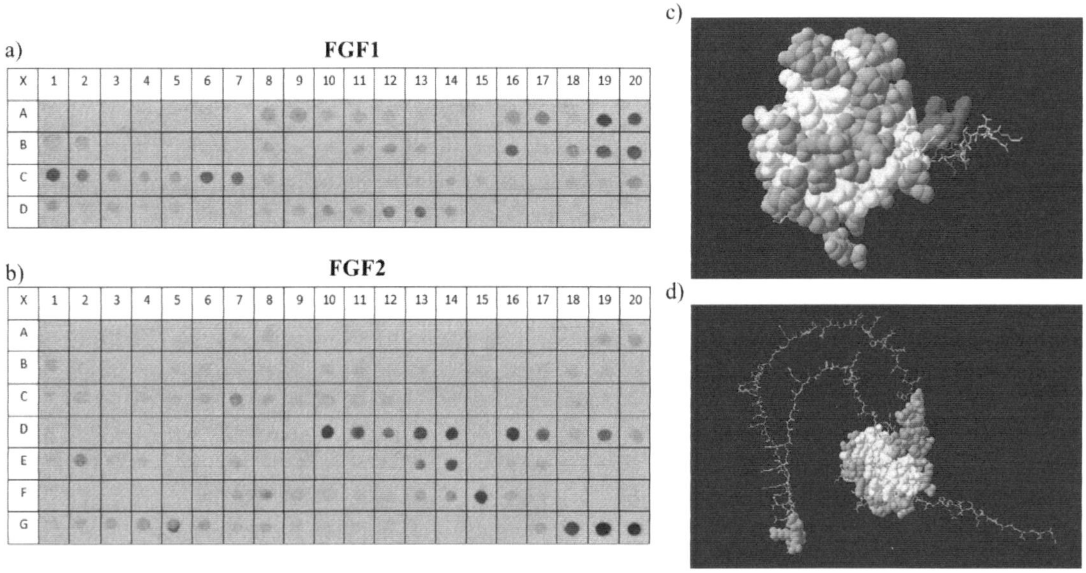

Fig. 1. a) Scan of a cellulose sheet containing the whole library of FGF1 fragments after reaction with antibodies. FGF1 map: 74 decapeptides, reading frame: 2 AA. b) Scan of a cellulose sheet containing the whole library of FGF2 fragments after reaction with antibodies. FGF2 map: 140 decapeptides, reading frame: 2 AA. c) FGF1 found fragments, AlphaFold model (Uniprot P05230). d) FGF2 found fragments, AlphaFold model (Uniprot P09038).

In the case of the FGF1 fragment library, 37 fragments able to interact with polyclonal antibodies were found based on the Dot-blot assay, including 17 fragments showing strong/very strong interaction. For the FGF2 fragment library, 20 fragments able to interact with polyclonal antibodies were found, of which 14 fragments showed strong/very strong interaction. The found fragments of FGF1 and FGF2 were synthesized on 2-chlorochlorotrityl resin using a triazine coupling reagent [13]. In the next step, FGF fragments were used in studies on NIH/3T3 cells to investigate their effect on the viability of cells incubated in a medium with the addition of FGF fragments.

http://doi.org/10.17952/37EPS.2024.P1116

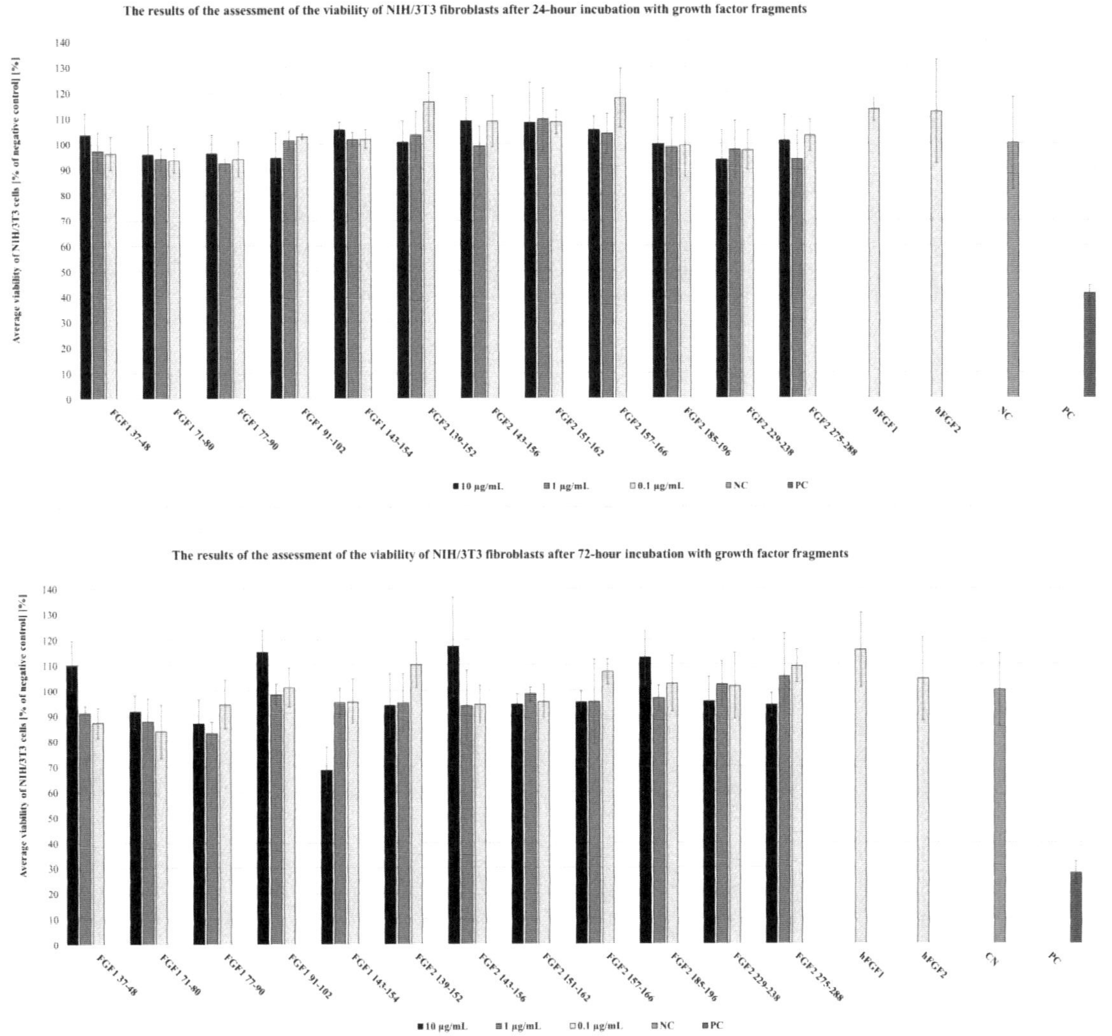

Fig. 2 Viability of NIH/3T3 cells grown in the presence of FGF1 and FGF2 fragments. Incubation for 24 h (top panel), incubation for 72 h (bottom panel).

Studies have shown that all FGF1 and FGF2 fragments have no negative effect on cell viability at concentrations of 1 and 0.1 µg/mL. However, at a concentration of 10 µg/mL, only one FGF1 fragment (cultured for 72 hours) showed a slight decrease in cell viability.

Conclusions

Selected FGFs fragments are useful in the regeneration of skin damage and treatment of difficult-to-heal wounds.

Literature

[1] Wobma, H., Vunjak-Novakovic, G. (2016) Tissue Engineering and Regenerative Medicine 2015: A Year in Review. *Tissue Eng. Part B Rev. 22*, 101-113. doi: 10.1089/ten.TEB.2015.0535.

[2] Marx, V. (2015) Organs from the lab *Nature 522*, 373-377. doi.org/10.1038/522373a.

[3] Masuki, H., Okudera, T., Watanebe, T., Suzuki, M., Nishiyama, K., Okudera, H. (2016) Growth factor and pro-inflammatory cytokine contents in platelet-rich plasma (PRP), plasma rich in growth factors (PRGF), advanced platelet-rich fibrin (A-PRF), and concentrated growth factors (CGF). *Int. J. Implant Dent. 2*, 19. doi.org/10.1186/s40729-016-0052-4.

[4] Giannotti, L., Di Chiara Stanca, B., Spedicato, F., Nitti, P., Damiano, F., Demitri, C., Calabriso, N., Carluccio, M.A., Palermo, A., Siculella, L. (2023) Progress in Regenerative Medicine: Exploring Autologous Platelet Concentrates and Their Clinical Applications. *Genes 14*, 1669. doi.org/10.3390/genes14091669.

 http://doi.org/10.17952/37EPS.2024.P1116

[5] Yahata, Y., Handa, K., Ohkura, N., Okamoto, M., Ohshima, J., Itoh, S., Kawashima, N., Tanaka, T., Sato, N., Noiri, Y., Hayashi, M., Okiji, T., Saito, M. (2023) Autologous concentrated growth factor mediated accelerated bone healing in root-end microsurgery: A multicenter randomized clinical trial. *Regen. Ther. 24*, 377-384. doi.org/10.1016/j.reth.2023.08.006.

[6] Rodella, L.F., Favero, G., Boninsegna, R., Buffoli, B., Labanca, M., Scarì, G., Sacco, L., Batani, T., Rezzani, R. (2011) Growth factors, CD34 positive cells, and fibrin network analysis in concentrated growth factors fraction. *Microsc. Res. Tech. 74*, 772-777. doi: 10.1002/jemt.20968.

[7] Powers, C.J., McLeskey, S.W., Wellstein. A. (2000) Fibroblast growth factors, their receptors and signaling. *Endocr. Relat. Cancer 7*, 165-197. doi:10.1677/erc.0.0070165.

[8] Sogabe, Y., Abe, M., Yokoyama, Y., Ishikawa, O. (2006)Basic fibroblast growth factor stimulates human keratinocyte motility by Rac activation. *Wound Repair Regen. 14*, 457-462. doi:10.1111/j.1743-6109.2006.00143.x

[9] Robson, M.C. (1997)The role of growth factors in the healing of chronic wounds. *Wound Repair Regen. 5*, 12-17. doi:10.1046/j.1524-475X.1997.50106.x

[10] Payne, W.G., Ochs, D.E., Meltzer, D.D., Hill, D.P., Mannari, R.J., Robson, L.E., Robson, M.C (2001) Long-term outcome study of growth factor-treated pressure ulcers. *Am J Surg. 181*, 81-86. doi:10.1016/S0002-9610(00)00536-5.

[11] Uchi, H., Igarashi, A., Urabe, K., Koga, T., Nakayama, J., Kawamori, R., Tamaki, K., Hirakata, H., Ohura, T., Furue, M. (2009) Clinical efficacy of basic fibroblast growth factor (bFGF) for diabetic ulcer. *Eur. J. Dermatology. 19*, 461-468. doi:10.1684/ejd.2009.0750.

[12] Fraczyk, J., Walczak, M., Kaminski, Z.J. (2018) New methodology for automated SPOT synthesis of peptides on cellulose using 1,3,5-triazine derivatives as linkers and as coupling reagents. *J. Pept. Sci., 24*, e3063. doi.org/10.1002/psc.3063.

[13] Kolesinska, B., Rozniakowski, K.K., Fraczyk, J., Relich, I., Papini, A.M., Kaminski, Z.J. (2015) The Effect of Counterion and Tertiary Amine on the Efficiency of N-Triazinylammonium Sulfonates in Solution and Solid-Phase Peptide Synthesis. *Eur. J. Org. Chem.*, 401-408. doi: 10.1002/ejoc.201402862.

Proceedings of the 37th European Peptide Symposium
Michal Lebl, Editor
European Peptide Society, 2024
http://doi.org/10.17952/37EPS.2024.P1138

Impaired metabolic phenotyping in mice with deletions of GPR10 and NPFFR2 receptors: Implications for lipidized PrRP analog therapy in obesity

Veronika Strnadová[1], Magdalena Škrlová[1], Jaroslav Kuneš[1,2], Lenka Maletínská[1]

[1] *Institute of Organic Chemistry and Biochemistry of the Czech Academy of Sciences, 16600 Prague, Czech Republic*
[2] *Institute of Physiology of the Czech Academy of Sciences, 14200 Prague, Czech Republic*

Scientific background: Obesity is a growing global health concern with limited treatment options. Prolactin-releasing peptide (PrRP) is a neuropeptide that can reduce food intake when administered centrally but loses this effect when given peripherally. However, lipidization of peptides can enhance their stability and efficacy after peripheral administration. Our previous work showed that lipidized PrRP analogs effectively reduced food intake and body weight in mice[1]. PrRP binds to its GPR10 receptor and with high affinity to neuropeptide FF receptor type 2 (NPFFR2)[2], both of which are involved in regulating food intake and energy homeostasis.

Palmitoylated analog of PrRP31, palm[11]-PrRP31 improved its binding and agonist activity for both receptors[3,6,7]. We have previously shown that deletion of GPR10 resulted in increased food intake, elevated insulin levels, impaired glucose tolerance, and altered lipid metabolism gene expression[4]. NPFFR2 knockout mice showed severe glucose intolerance, which worsened with a high-fat diet (HFD), and developed hypothalamic insulin resistance due to reduced insulin pathway signaling proteins[5]. Deletion of both GPR10 and NPFFR2 caused sex-specific and diet-dependent metabolic changes, including impaired glucose tolerance and hyperglycemia on a high-fat diet.

Methods: In this study, we examined the effect of palm[11]-PrRP31 on mice with deletion of either GPR10 or NPFFR2 or both receptors (GPR10/NPFFR2 KO or dKO mice). Male mice were fed a HFD starting at 8 weeks of age for a duration of 5 months. During the final month, they received a 4-week subcutaneous treatment with either saline or palm[11]-PrRP31 (5 mg/kg) administered twice daily. An oral glucose tolerance test (OGTT) was performed in the third week of the interventions, and fasting plasma was collected to measure insulin levels. Finally, the mice were euthanized by decapitation, and organs (brain, liver, adipose tissue) were collected for further analysis.

Results and discussion: The palmitoylated palm11-PrRP31 analog demonstrated strong appetite-suppressing and glucose-lowering effects in obese wild-type (WT) mice. Treatment with palm11-PrRP31 led to a significant reduction in body weight and food intake in GPR10 KO mice, whereas in NPFFR2 KO mice, the reduction in body weight was less pronounced and food intake was only slightly reduced. In dKO mice, which lack both GPR10 and NPFFR2 receptors, there was no reduction in body weight or food intake.

http://doi.org/10.17952/37EPS.2024.P1138

Palm[11]-PrRP31 significantly improved glucose tolerance in WT mice, but had no effect in any of the KO models, suggesting that the activation of both GPR10 and NPFFR2 receptors is essential for its anti-obesity and anti-diabetic effects.

Previous studies have shown that the GPR10 and NPFFR2 receptors are involved in regulating energy balance, food intake, and glucose homeostasis. The dual agonism of these receptors by palm[11]-PrRP31 may enhance anorectic signaling pathways, reduce insulin resistance, and promote glucose uptake in peripheral tissues, which are critical mechanisms for weight loss and improved metabolic control. Moreover, lipidization of PrRP analogs, such as palm[11]-PrRP31, increases their stability and prolongs their half-life, enhancing their therapeutic efficacy. Therefore, targeting both GPR10 and NPFFR2 receptors with lipidized PrRP analogs could offer a promising strategy for treating obesity and related metabolic disorders by addressing both appetite regulation and glucose metabolism, potentially overcoming the limitations of current therapies that target only a single pathway.

Dedication:

This research was supported by the project National Institute for Research of Metabolic and Cardiovascular Diseases (Programme EXCELES, ID Project No. LX22NPO5104) - Funded by the European Union - Next Generation EU, by grant 21-03691S Czech Science Foundation of the GACR and by RVO: 61388963 and RVO:67985823 of the CAS.

References:

1. Maletínská L. et al., Int J Obes (2015), 39, 986–993
2. Engström M. et al., J Pharmacol Exp Ther (2003), 305(3):825-32
3. Pražienková V., Holubová M. et al.,PLoS ONE 12(8): e0183449
4. PražienkováV. et al., Gene (2021), 774-145427
5. Karnošová A., Strnadová V. et al., Clin Sci (2023), 137 847–862
6. Strnadová V. et al., Neuropeptides 98 (2023), 102319
7. Karnošová A. et al., Int J Mol Sci (2021), 22, 8904

Proceedings of the 37th European Peptide Symposium
Michal Lebl, Editor
European Peptide Society, 2024
http://doi.org/10.17952/37EPS.2024.P1149

Synthesis and Characterization of Tumor Targeted Peptide (TTP) Conjugated with Magnetic Nanoparticles for Thermo-ablation

Sima Alvani Alamdari[1], Justyna Fraczyk[1], Joanna Waśko[1], Anita Plazinska[2], Beata Kolesinska[1]

[1]Institute of Organic Chemistry, Faculty of Chemistry, Lodz University of Technology, Zeromskiego 116, 90-924 Lodz, Poland.
[2]Department of Biopharmacy, Faculty of Pharmacy, Medical University of Lublin, Chodźki 4a, 20-093 Lublin, Poland.
Email: sima.alvani-alamdari@dokt.p.lodz.pl

Introduction :

Nanotechnology is transforming various industries by enabling the manipulation of materials at the nanometer scale, leading to breakthroughs in drug delivery, water purification, data management, and the creation of advanced nanoscale materials for healthcare and industrial applications. By leveraging the unique properties of nanometric materials—such as their small size, large surface area, and customizable surfaces—scientists are developing innovative solutions with significant potential, particularly in targeted drug delivery [1-3].

Magnetic nanoparticles (MNPs), especially superparamagnetic iron oxide nanoparticles (SPIONs), have emerged as key players in biomedical research due to their promising applications in cancer treatment and other medical fields. These nanoparticles can be precisely engineered to enhance the delivery of therapeutic agents directly to cancer cells, reducing side effects and improving treatment efficacy.

Peptides are an attractive alternative to antibody-based targeted therapies. Unlike antibodies, peptides are easy to synthesize in large quantities, and their smaller size improves tissue penetration while preventing nonspecific uptake by the reticuloendothelial system. In addition, peptides can be chemically modified to alter affinity, charge, hydrophobicity, stability, and solubility, and can be optimized for in vivo use through modification. Importantly, peptides can exhibit receptor affinity similar to antibodies. Tumor-targeting peptides (TTPs) are an effective alternative for selectively targeting specific tumor receptors. Compared to antibodies, TTPs have better tumor/tissue penetration and are easy to synthesize and chemically modify to improve stability and pharmacokinetics. Peptides can be used for targeted drug delivery. They can be designed to have the ability to destroy cancer cells and inhibit tumor growth. Examples of TPPs are [4]: RGD-4C, iRGD, NGR, which act on: B16 melanoma, MDA-MB-435, HL-60, MMTV-PyMT, KRIB, C8161, K14-HPV16/E2 RIP1-Tag2, K14-HPV16.

This project focuses on synthesizing and characterizing novel hybrid nanomaterials (MNP-CPP-TM-drug) that combine MNPs with active anti-cancer compounds (EGF fragments connected with cytostatic). The goal is to develop a multifunctional nanostructure capable of not only delivering drugs with high specificity but also enabling targeted thermo-ablation for enhanced cancer.

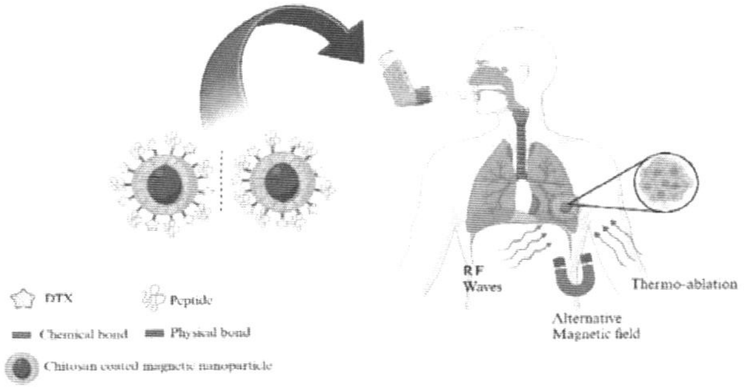

 http://doi.org/10.17952/37EPS.2024.P1149

Methodology :

In the first stage of the research, magnetic nanoparticles (size: 14-29 nm, 99.5%) were coated with chitosan (low molecular weight) physically and chemically modified by different cross-linkers. Chitosan coating is biodegradable and helps attach other elements, such as anticancer drugs and targeting peptides by cross-linker. Targeting molecules were EGF fragments capable of interacting with the EGF receptor (EGFR). Docetaxel was used as an anticancer drug in this research.
EGF fragments in this research were prepared by solid-phase peptide synthesis (SPPS) with Fmoc/tBut strategy using triazine coupling reagent (4-(4,6-dimethoxy-1,3,5-triazin-2-yl) -4-methylmorpholinium toluene-4-sulfonate (DMT/NMM/TosO-))[5]. All EGF fragments were characterized by a high ability to interact with EGFR (microscale thermophoresis (MST)) [6].

Results and discussions :

Nanoparticles were characterized with SEM, DLS, FTIR, and TGA. Additionally, the peptides were characterized by LC-MS.

Table 1. LC-MS analysis of EGF fragments.

peptide	Sequence	Molecular wight	M/Z	HPLC Purity%
EGF fragment	^{10}HDGYCL^{15}	706.263	$[M+H]^{1+}$=707.2867 $[M+2H]^{2+}$=354.1477	96%
EGF fragment	^{43}QYRDL^{47}	693.334	$[M+H]^{1+}$=694.3590 $[M+2H]^{2+}$=347.6831	90%

Comparing coated magnetic nanoparticles with different cross-linkers: Dynamic light scattering (DLS) measurement was used to compare the sizes of MNPs (table 2), and simple Acid-base titration was used to evaluate the number of primary amine groups on the surface of MNPs (table 3).

Table 2. DLS measurement of MNPs (pH=7).

No.	MNPs	Size (nm)
1	Fe_3O_4@CS(300CPS)/TPP	70-71
2	Fe_3O_4@CS(30CPS)/TPP	55
3	Fe_3O_4@CS/Citric acid	90-130
4	Fe_3O_4@CS/Squaric acid	105
5	Fe_3O_4@CS/glutaraldehyde	253-263

Table 3. Result from Acid-base titration method.

sample	Cross-linker	$V_{ave. / mL}$	NH_2 group (mol /mg)
1	TPP	1.73	0.293
2	Citric Acid	1.96	0.383
3	Squaric acid	1.85	0.272
4	Glutaraldehyde	1.63	0.065

Magnetic nanoparticles were also characterized by scanning electron microscopy (SEM), thermogravimetric analysis (TGA), and Fourier transform infrared spectroscopy (FT-IR) (Figure 1).

http://doi.org/10.17952/37EPS.2024.P1149

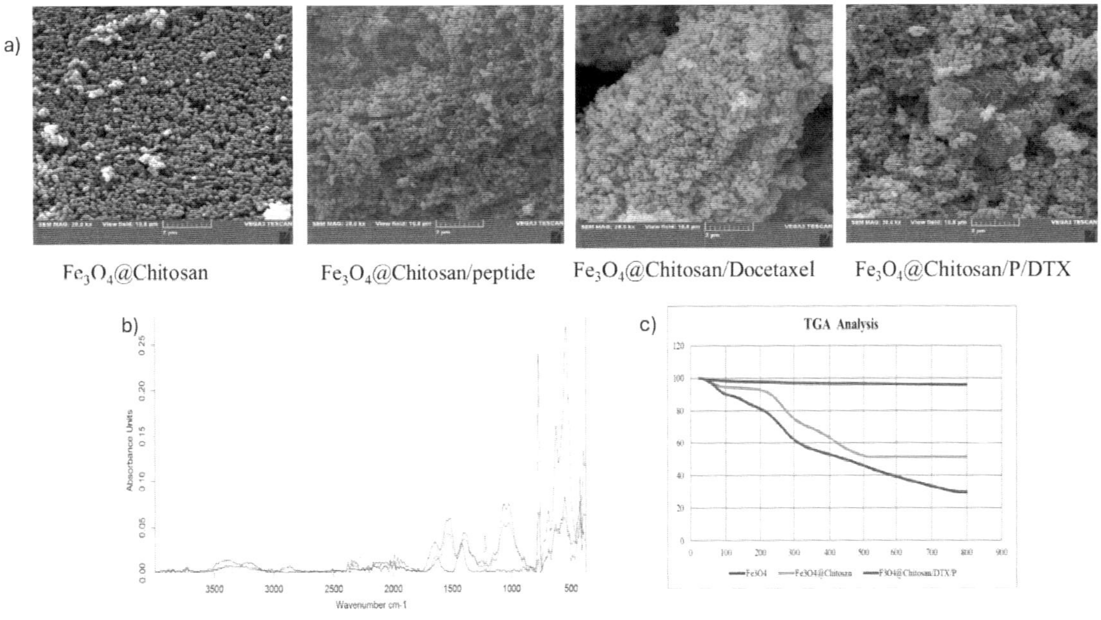

Figure 1. a) SEM images of chitosan-coated magnetic nanoparticles, b) FT-IR spectra of 1)Fe3O4, 2)Fe3O4@Chitosan, 3)Fe3O4@Chitosan/DTX/P, and c) TGA spectra of synthesized nanoparticles

Conclusion :

In this research, an effective method for coating magnetic nanoparticles with chitosan was developed. Tripolyphosphate (TPP) was chosen as the optimal cross-linker for synthesizing these coated MNPs. In the second step, an anticancer drug (DTX) and a peptide derived from EGF fragments were attached to the MNPs both physically and chemically. The thermo-ablation properties of these nanoparticles were also studied, and they are currently in their preliminary phase of research.

The nanoparticles were characterized using SEM, DLS, FTIR, and TGA. Additionally, peptides were characterized by LC-MS. The biological activity of the tumor-targeting peptide conjugated with magnetic nanoparticles and the cytostatic agent was confirmed.

Acknowledgments:

Research reported in this publication is supported by the "FU²N - Fund for Improving the Skills of Young Scientists" program, internal grants were awarded in the discipline of Chemical Sciences under grant No. 503-3-32-4-1 from the Lodz University of Technology.

References :

[1] Zhou, L. et al., J. Mater. Chem. 2011,21, 9, 2823-40.
[2] Ziegler-Borowska, M. et al., Catalysts. 2017, 7, 1, 26.
[3] Azarmi, S.et al., Adv. Drug Deliv. Rev. 2008, 60, 8, 863-75.
[4] Simón-Gracia, L. et al., Molecules, 2018, 23, 1190.
[4] Kamiński, Z.J. et al., J. Am. Chem. Soc. 2005, 127, 48, 16912-20.
[5] Czerczak-Kwiatkowska, K. et al. Int. J. Mol. Sci. 2024, 25, 3, 1470.

Proceedings of the 37th European Peptide Symposium
Michal Lebl, Editor
European Peptide Society, 2024 http://doi.org/10.17952/37EPS.2024.P1150

Synthesis and characterization of phenethyl isothiocyanate conjugated with magnetic nanoparticles modified by tumor targeting peptide (TTP) using in cancer treatment

Sima Alvani Alamdari[1], Łukasz Janczewski[1], Justyna Fraczyk[1]
[1]*Institute of Organic Chemistry, Faculty of Chemistry, Lodz University of Technology, Zeromskiego 116, 90-924 Lodz, Poland.*
Email: sima.alvani-alamdari@dokt.p.lodz.pl

Introduction:

Many isothiocyanates, both natural and synthetic, show anticarcinogenic activity because they decrease activation of carcinogens and increase their detoxification. They exhibit anti-tumor activity. [1].

One of the greatest challenges in cancer therapy is to selectively deliver chemotherapeutics to the cancer cells, thus achieving therapeutic efficacy while limiting damage to healthy tissues. The effective strategy of targeted drug delivery to tumor tissues is based on cellular heterogeneity which is characteristic of all organs and tissues including pathological tissues such as tumors. One of the biological properties of tumor cells is the expression or overexpression of a unique set of proteins or receptors on the cell surface. Activation of these receptors often results in deregulated signaling, increased cell proliferation and decreased cell apoptosis. Selective targeting of these tumor-specific receptors provides the opportunity for targeted drug delivery. Several types of tumor targeting ligands have been developed including antibodies, antibody fragments, polypeptides, small molecules, and others [2]. Tumor-targeting peptides (TTPs, targeting peptide (TM), homing peptide) are an efficient alternative for selective targeting of tumor specific receptors. Compared to antibodies, tumor-targeting peptides have better tumor/tissue penetration and are easy to synthesize and modify chemically to improve stability and pharmacokinetics. The first generation tumor homing peptides, RGD and NGR, have entered clinical trials. As a molecular targets of NSCLC, have been investigated: epidermal growth factor receptor (EGFR), Kirsten rat sarcoma viral oncogene homolog (KRAS), anaplastic lymphoma kinase (ALK), human epidermal growth factor receptor 2 (HER2) [3]. EGFR, one of these molecular targets with a high frequency of mutation, is a transmembrane receptor tyrosine kinase involved in the signaling pathways regulating cell proliferation, apoptosis, angiogenesis, and invasion [4].

In this study, we report the synthesis and characterization of phenethyl isothiocyanate (PEITC) conjugated onto the surface of magnetic nanoparticles (MNPs) modified by a tumor-targeting peptide (TTP) for enhanced cancer treatment and to facilitate targeted drug delivery [5]. PEITC, , which is one of the most studied ITCs, shows anti-cancer and chemopreventive properties and inhibits the growth of cancer cells by inducing apoptosis and inhibiting the cell cycle, making it an attractive candidate for cancer therapy. Additionally, TTP was incorporated to improve specificity towards cancer cells.

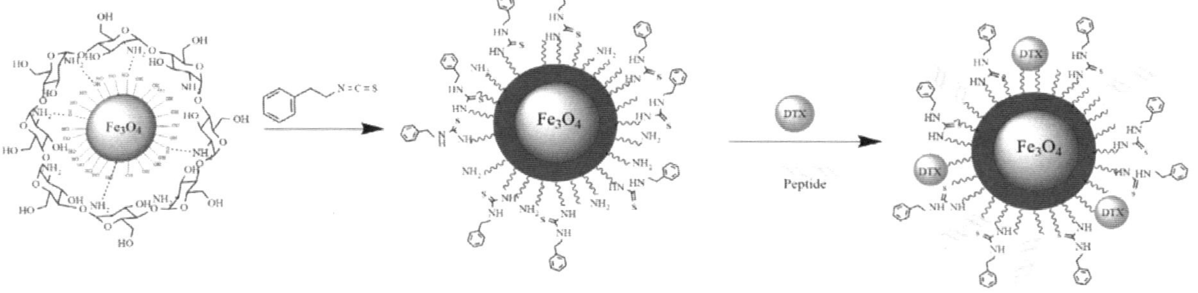

 http://doi.org/10.17952/37EPS.2024.P1150

Moreover, MNPs offer several advantages such as biocompatibility, ease of functionalization, and the ability to be guided to specific sites using an external magnetic field. By conjugating PEITC onto MNPs, we aimed to enhance its delivery to cancer cells while minimizing off-target effects.

Methodology:

In this synthesis method, phenethyl isothiocyanate (PEITC) is chemically attached to the chitosan (low molecular weight) using the method described by Bae et al.'s method [6]. Then magnetic nanoparticles (size: 14-29 nm, 99.5%) were coated by phenethyl isothiocyanate conjugated chitosan. Following PEITC conjugation, a tumor-targeting peptide (TTP) is incorporated onto the surface of the nanoparticles to impart specificity to cancer cells. TTP were EGF fragments capable of interacting with the EGF receptor (EGFR) [7]. Additionally, Docetaxel was used as an anti-cancer drug.

EGF fragments in this research were prepared by solid-phase peptide synthesis (SPPS) with Fmoc/t-Bu strategy using triazine coupling reagent (4-(4,6-dimethoxy-1,3,5-triazin-2-yl) -4-methylmorpholinium toluene-4-sulfonate (DMT/NMM/TosO$^-$)) [8].

Results and discussions :

Nanoparticles were characterized with TEM, DLS, FTIR, and TGA, and the structure of chitosan modified with PEITC was confirmed with NMR and FT-IR (Figures 1 and 2). Additionally, the peptides were characterized by LC-MS.

Characterization of modified chitosan :

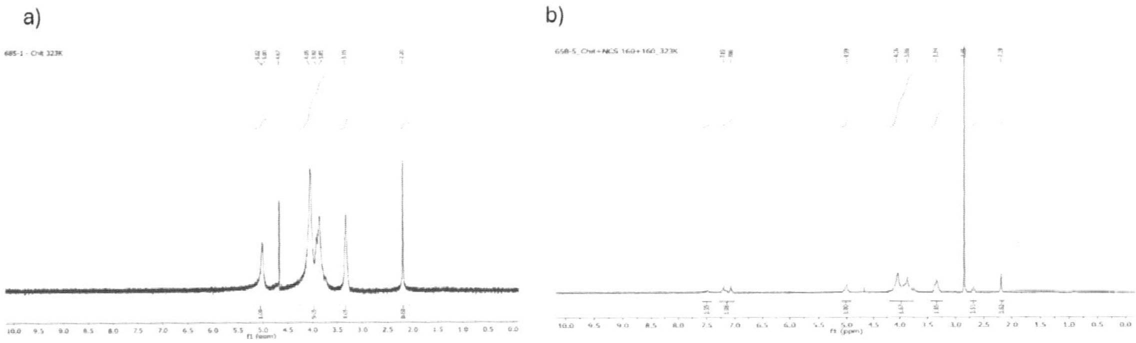

Fig. 1 ^{1}H NMR NMR (700 MHz, D_2O) spectra's of a) Free chitosan (CS) and b) Modified chitosan with PEITC.

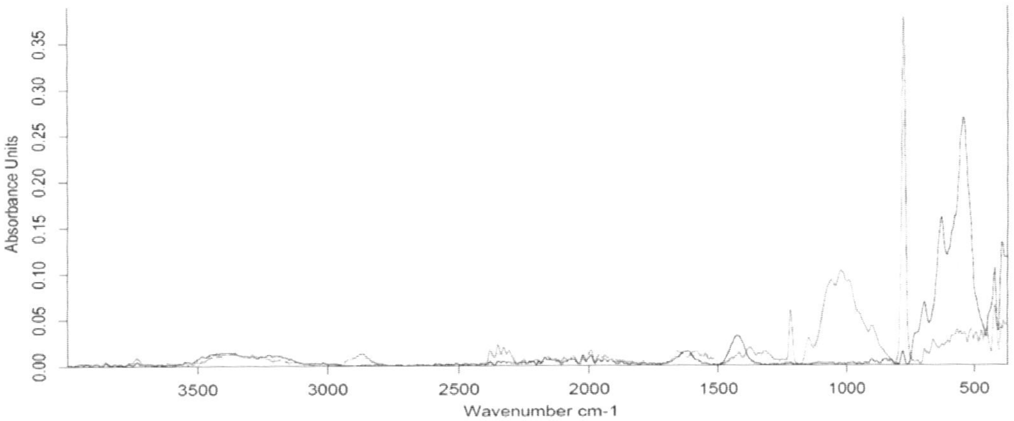

Fig. 2 FT-IR spectra's of a) Fe_3O_4, b) Chitosan(CS), and c) Fe_3O_4@CS/PEITC.

 http://doi.org/10.17952/37EPS.2024.P1150

Characterization of peptides by LC-MS :

Table 1. LC-MS analysis of EGF fragments.

Peptide	Sequence	Molecular weight [g/mol]	m/z	HPLC Purity [%]
EGF fragment	[10]HDGYCL[15]	706.263	$[M+H]^{1+}$=707.2867 $[M+2H]^{2+}$=354.1477	96%
EGF fragment	[43]QYRDL[47]	693.334	$[M+H]^{1+}$=694.3590 $[M+2H]^{2+}$=347.6831	90%

Characterization of chitosan-coated MNPs modified with peptide and anti-cancer drug:

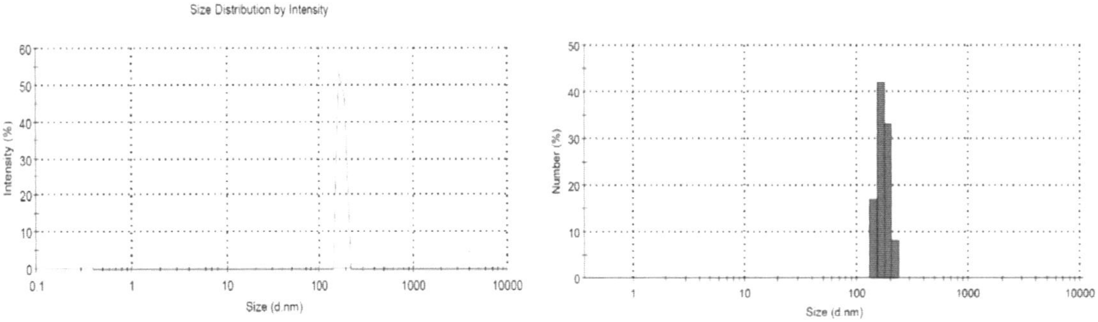

Fig. 3 DLS spectra of Fe_3O_4@chitosan/PEITC (Ave. size: 176 nm).

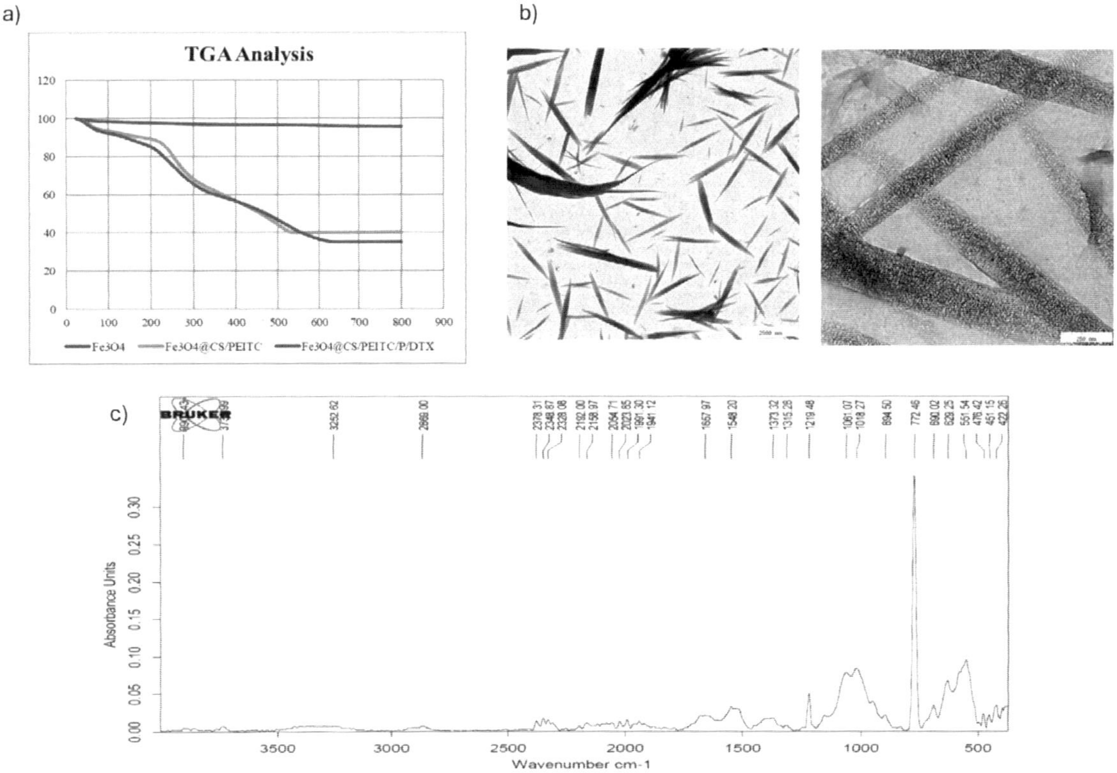

Fig. 4 a) TGA spectra of synthesized nanoparticles, b) TEM images of Fe_3O_4@chitosan/PEITC, and c) FT-IR spectra of Fe_3O_4@chitosan/PEITC/DTX/Peptide.

 http://doi.org/10.17952/37EPS.2024.P1150

Conclusion:

In this project, an effective method for synthesizing phenethyl isothiocyanate conjugated with magnetic nanoparticles was developed. In the second step, an anticancer drug (DTX) and a peptide derived from EGF fragments were attached to the MNPs both physically and chemically interaction. The nanoparticles were characterized using TEM, NMR, DLS, FTIR, and TGA. Additionally, peptides were characterized by LC-MS, and their structures were confirmed. The interaction between peptides and MNPs will be studied.

References:

[1] Wu, X. et al., *Acta Pharmacol. Sin.* 2009. *30*, 501–512.
[2] Raucher, D. *Curr. Opin. Pharmacol.* 2019, *47*, 14-19.
[3] Cheng, Z et al., *BMC Med. Imaging* 2017, *17*, 5.
[4] Nishino, M.et al., *Radiology*, 2014, *271*, 6-27.
[5] Laurent, S. et al., *Chem. Rev.* 2008, *108*, 2064-2110.
[6] Bae, I. et al., *Int. J. Mol. Sci.* 2022, *23*, 13802.
[7] Czerczak-Kwiatkowska, K. et al., *Int. J. Mol. Sci.* 2024, 25, 3, 1470.
[8] Kamiński, Z.J. et al., *J. Am. Chem. Soc.* 2005, 127, 48, 16912-20.

Proceedings of the 37 th European Peptide Symposium
Michal Lebl (Editor)
European Peptide Society, 2024
https://doi.org/10.17952/37EPS.2024.P1162

Peptide-based vesicle isolation method for *P. aeruginosa*

Tania Henriquez, Ismael Castañón, Silvia Scali, Chiara Falciani.
Department of Medical Biotechnologies, University of Siena, Siena, Italy.

Introduction

Pseudomonas aeruginosa is an organism that can act as an opportunistic pathogen for humans and animals, where it can cause from mild diseases, such as otitis, to serious illnesses (life-threatening diseases) such as septicemia (1).

In the last decades, new elements have been associated with the pathogenicity of *Pseudomonas* strains (2). This is the case with outer membrane vesicles (OMVs), which are produced when part of the outer membrane (including periplasmic content) generates a bleb. Bacterial vesicles have diverse roles in physiology and virulence (3). However, the difficulties related to vesicle purification (i.e. long protocols or the need of specialized instruments) have become a major obstacle for their characterization. In this context, we decided to analyze the efficacy of a peptide designed in our lab to bind *P. aeruginosa* vesicles and use it to develop a new vesicle purification method.

Results and discussion

We first determine the best conditions to purify vesicles of *P. aeruginosa* PAO1 strain using a commercial kit (ExoBacteria OMV isolation kit) (Figure 1). To that end, we grew the strain in rich medium (KB) or M9 minimal medium. In parallel, we also used KB medium and M9 minimal medium without bacteria as negative control.

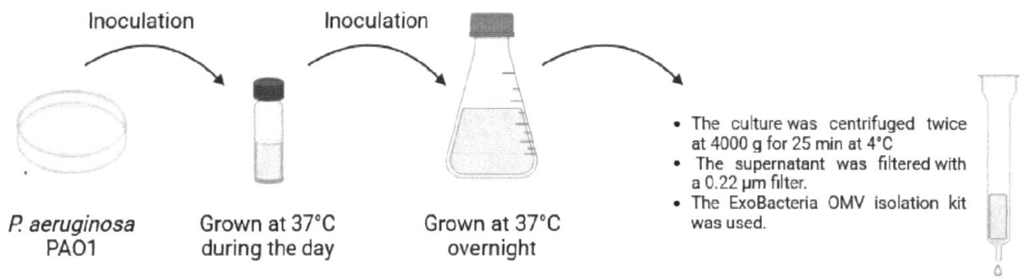

Figure 1. Diagram showing the main steps for vesicle purification. P. aeruginosa *PAO1 was grown in a cetrimide plate and then used to inoculate 2 ml of the corresponding medium. The strain was grown at 37°C during the day, and then used to inoculate 30 of fresh medium. After overnight incubation at 37°C and continuous shaking, the culture was centrifuged twice, filtered (0.25 µm) and processed with the ExoBacteria OMV Isolation according to the manufacturer instructions. Image created with BioRender.com*

We proceeded to characterize all samples—including those from *P. aeruginosa* cultures and negative controls in KB and M9 minimal media—using dynamic light scattering, electron microscopy, and protein quantification assays. Our findings demonstrated that the M9 minimal medium was the most effective for vesicle purification with the ExoBacteria OMV isolation kit. The negative control in M9 minimal medium exhibited a profile consistent with the absence of particle populations, whereas the negative control in KB medium (without bacteria) showed the presence of particle populations, likely due to protein aggregates (figure not shown). Furthermore, our analysis confirmed the presence of *P. aeruginosa* vesicles in the M9 minimal medium (Figure 2). Consequently, we selected these vesicles for subsequent experiments.

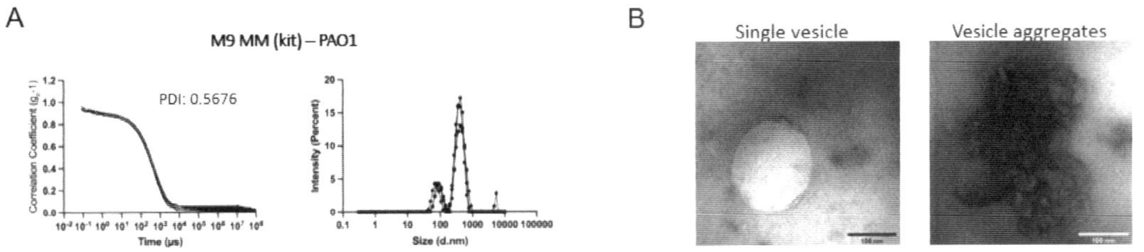

Figure 2. ExoBacteria OMV Isolation Kit was used to purify vesicles/particles from M9 minimal medium inoculated with P. aeruginosa *PAO1 strain. (A) Analysis of growth media for vesicle purification by Dynamic Light Scattering (DLS) using DLS (Zetasizer). The three curves in each graph represent the three readings of the instrument (triplicate). PDI: polydispersity index. (B) Electron microscopy images.*

We later synthesized an amphipathic and cationic 9-mer peptide that was previously described to bind LPS of *Escherichia coli* and *P. aeruginosa* (4) and used it to perform a modified Differential Radial Capillary Action of Ligand Assay (DraCaLa) assay (5). This DraCaLa assay relied on the colorimetric properties of our peptide conjugated to tetramethylrhodamine for visualization of migration on the nitrocellulose membrane (instead of the traditional radiolabeling). Our results suggested that the 9-mer peptide could bind the vesicles of *P. aeruginosa* PAO1. Interestingly, similar results were found for human exosomes (although with lower affinity).

Additional experiments are needed in order to determine the binding affinity between the peptide and the vesicles. In this context, ELISA experiments will be performed.

Altogether, these results contribute to broadening the knowledge on *Pseudomonas* vesicle purification and generate a new alternative for future applications.

Acknowledgments

We thank Eugenio Paccagnini, Mariangela Gentile and Pietro Lupetti for their help with Electron microscopy. Also, we thank Ilaria Clemente, Claudia Bonechi and Agnese Magnani for their help with the dynamic light scatter analysis.

References

(1) Iglewski BH. *Pseudomonas*. In: th, Baron S, editors. Medical Microbiology. Galveston TX: The University of Texas Medical Branch at Galveston.; 1996.

(2) Palleroni NJ. Pseudomonas. Bergey's Manual of Systematics of Archaea and Bacteria. 2015.

(3) Henriquez T, Falciani C. Extracellular Vesicles of *Pseudomonas:* Friends and Foes. Antibiotics 2023, 12, 703. https://doi.org/10.3390/antibiotics12040703.

(4) Falciani, Chiara & Lozzi, Luisa & Pollini, Simona & Luca, Vincenzo & Carnicelli, Veronica & Brunetti, Jlenia & Lelli, Barbara & Bindi, Stefano & Scali, Silvia & Di Giulio, Antonio & Rossolini, Gian & Mangoni, Maria & Bracci, Luisa & Pini, Alessandro. (2012). Isomerization of an Antimicrobial Peptide Broadens Antimicrobial Spectrum to Gram-Positive Bacterial Pathogens. PloS one. 7. e46259. 10.1371/journal.pone.0046259.

(5) Seminara AB, Turdiev A, Turdiev H, Lee VT. Differential Radial Capillary Action of Ligand Assay (DRaCALA). Curr Protoc Mol Biol. 2019 Apr;126(1):e84. doi: 10.1002/cpmb.84.

Proceedings of the 37th European Peptide Symposium
Michal Lebl, Editor
European Peptide Society, 2024
http://doi.org/10.17952/37EPS.2024.P1218

Development of Daunomycin-Peptide Antitumor Conjugates Containing Neurotensin-Based Homing Moiety

Levente Endre Dókus[1], Eszter Lajkó[3], Orsolya Láng[3], László Kőhidai[3], Gábor Mező[1,2]

[1]HUN-REN-ELTE Research Group of Peptide Chemistry, Budapest, Hungary.
[2]Eötvös Loránd University, Faculty of Science, Institute of Chemistry, Budapest, Hungary
[3]Department of Genetics, Cell- and Immunobiology, Semmelweis University, Budapest, Hungary.

Introduction

Pancreatic ductal adenocarcinoma (PDAC) is one of the most dangerous cancerous diseases of the pancreas leading to high mortality [1]. Targeted tumor therapy could be a good tool to treat PDAC efficiently. From the literature it is known that PDAC cells overexpress the neurotensin-1 (NTSr1) receptor. This protein binds to neurotensin, a peptide hormone consisting 13 amino acids, as its ligand therefore, it can be a potential target for neurotensin-based homing peptide drug conjugates. The biological activity of neurotensin is attributed to its C-terminal part: ^{8}RRPYIL^{13} [2]. Our aim was to develop new peptide - drug conjugates containing daunomycin as an anticancer agent attached to the homing peptides *via* an oxime bond [3].

Results and Discussion

In the first round six different daunomycin - peptide conjugates were initially developed according to the literature background. In the case of three conjugates, enzyme labile spacer (GFLG) [4] was built into the sequences between the homing peptide and the aminooxy moiety used for oxime ligation. This spacer is cleavable by the lysosomal enzyme cathepsin B, making it suitable for enhancing the intracellular degradation and the release of the active metabolite.

The aminooxyacetyl-group containing homing peptide parts were synthetized by SPPS using Fmoc/^{t}Bu technique on Wang-resin. The payload (Dau) was coupled in solution phase (0.2 M NH_4OAc solution (pH=5)) to the cargo *via* oxime bond (Figure 1). The cytotoxic effect of the conjugates were characterized on PANC-1 (a PDAC type) cell line by an impedimetric technique, xCELLigence System.

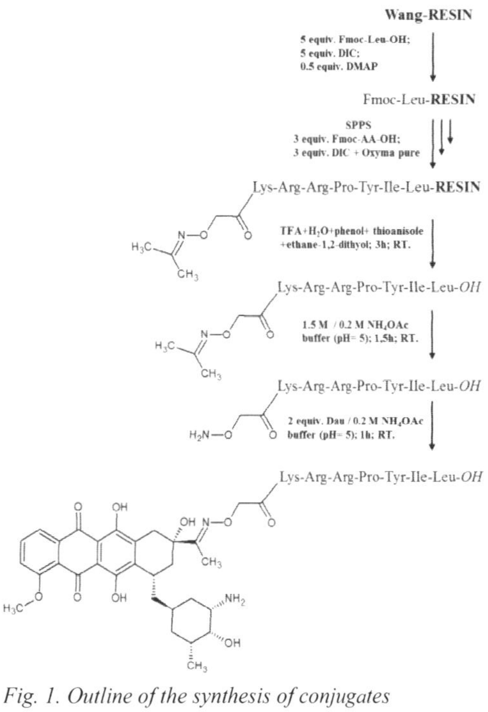

Fig. 1. Outline of the synthesis of conjugates

The results shown that from the first six constructions three conjugates (**2**,**5**,**6**) have a significant cytotoxic effect at 10^{-5} M concentration after 72 hours treatment, but the other three derivatives did not induce antitumor effect on PANC-1 cells (Table 1). The cellular uptake of these conjugates and their metabolism by lysosomal enzymes were also investigated (Figure 2 and 3).

These results were applied for design and prepare new non-natural amino acid (*e.g. tert*-leucine) containing derivatives to increase their enzyme stability [5]. Their antitumor effect (Table 1), internalization ability and lysosomal degradation were also investigated. (Figure 2 and 3)

In the next step the mentioned enzyme labile spacer was built in to the N-terminus of the two structures which resulted highest cellular uptake (**11** and **12**). Finally, these two structures with best internalization ability (**7**, **8**) were modified with additional non-natural amino acids and leaving of the *tert*-Leu. In the position^{8} of this two structures Arg(Me), *h*-Lys, *h*-Arg, D-Lys, D-Arg or in one case L-Arg were built in resulted nine new conjugates. The antitumor activity were also characterized on PANC-1 cell line in 10^{-5} M and 3.3×10^{-5}M concentrations using 72 or 96 hours incubation. Finally, the IC_{50} values of the three most efficient non-natural amino-acid containing derivatives were characterized (Table 1).

Table 1. Cytotoxic effect of the conjugates

	Sequence	Viability %		Sequence	Viability %	
1	*Dau*=*Aoa*-RRPYIL-*OH*	**> 50**	10	*Dau*=*Aoa*-KRKPW'LL-*OH*	102.6 ± 18.7	
2	*Dau*=*Aoa*-KRRPYIL-*OH*	4.7 ± 1.8	11	*Dau*=*Aoa*-GFLGKRRPY'LL-*OH*	116.4 ± 3.1	
3	*Dau*=*Aoa*-KPRRPYIL-*OH*	**> 50**	12	*Dau*=*Aoa*-GFLGKRRPW'LL-*OH*	116.6 ± 6.7	
4	*Dau*=*Aoa*-GFLGRRPYIL-*OH*	**> 50**		Sequence	IC_{50} (μM)	
5	*Dau*=*Aoa*-GFLGKRRPYIL-*OH*	13.0 ± 1.5			72 h	96 h
6	*Dau*=*Aoa*-GFLGKPRRPYIL-*OH*	11.0 ± 1.7	13	*Dau*=*Aoa*-K-*h*Arg-RPYIL-*OH*	27.8 ± 5.9	23.1 ± 14.2
7	*Dau*=*Aoa*-KRRPY'LL-*OH*	116.4 ± 3.1	14	*Dau*=*Aoa*-K-*h*Lys-RPWIL-*OH*	20.9 ± 2.3	13.1 ± 0.9
8	*Dau*=*Aoa*-KRRPW'LL-*OH*	116.6 ± 6.2	15	*Dau*=*Aoa*-*K*-Arg(Me)-RPYIL-*OH*	22.9 ± 2.5	16.7 ± 2.4
9	*Dau*=*Aoa*-KRKPY'LL-*OH*	110.0 ± 6.8	16	*Dau*=*Aoa*-K-D-Lys-RPWIL-*OH*	20.9 ± 2.2	20.3 ± 6.5

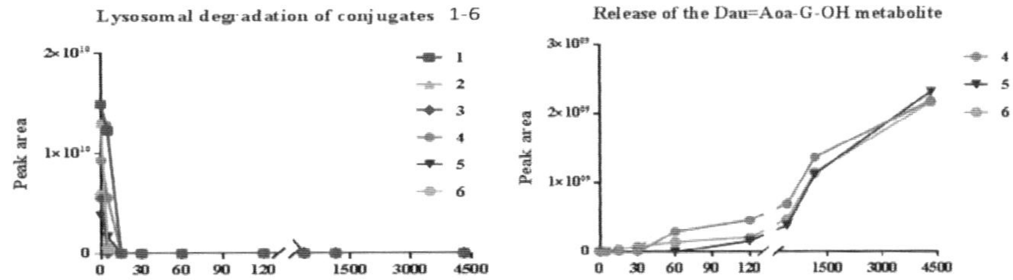

Fig. 2. Time dependent degradation of conjugates by lysosomal enzymes

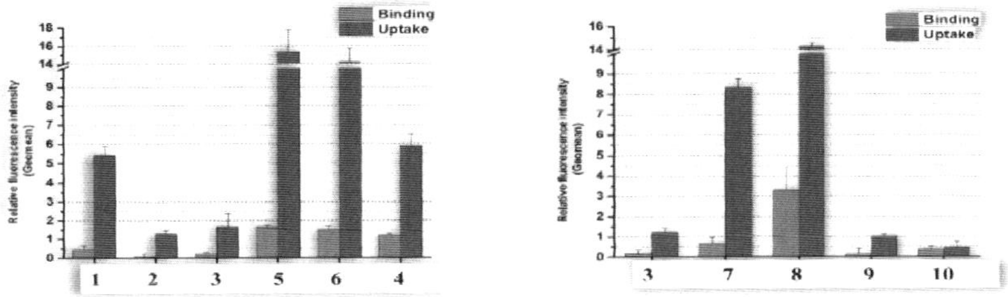

Fig. 3. Cellular binding and uptake of the conjugates

As discussion the results showed that three conjugates (**2, 5** and **6**) have significant cytotoxic effect at 10^{-5} M concentration (the cell (PANC-1) viability was lower than 15%). All of the investigated conjugates were fast degraded by lysosomal enzymes (within 30 minutes) but in case of GFLG containing constructions the presence of the smallest (active) metabolite in huge amount can be observed only after 60 minutes reaction time. (Figure 2) The presence of 'Leu in the 12 position increases the cellular uptake (Figure 3), but these conjugates did not present any antitumor effect (Table 1). The presence of the enzyme labile spacer GFLG does not increase significantly the effectivity in case of different homing moieties.The incorporation of non-native amino acids into the position 8 of homing peptide results high cytotoxicity in case of **13**, **14**, **15**, and **16** conjugates, but only in higher (33 μM) concentration.

These conjugates might be good candidates for further development (other drugs and/or linker). In conclusion, the set of the developed neurotensin derivative – drug conjugates provide evidences to select appropriate molecules for further therapeutic experiments or as diagnostic tools.

Acknowledgments

The project was supported by the National Research, Development and Innovation Office (NKFIH K146039 and NVKP_16-1-2016-0036)

References

[1]: Jemal, A. et al. (2007) *CA Cancer J. Clin.* **57**, 43-66.

[2]: Gilbert, J.A. et al. (1989) *Biochem. Pharm.* **38**, 3377-3382.

[3]: Mező, G. et al (2024) *Int. J. Mol. Sci.* **2024**, *25*(3), 1864.

[4]: Peterson, J.J. et al. (1998) *Bioconj. Chem.* **9**, 618–626.

[5]: Hadden, M. K. et al. (2005) *Neuropharm.* **49**,1149-1159.

Proceedings of the 37th European Peptide Symposium
Michal Lebl, Editor
European Peptide Society, 2024 http://doi.org/10.17952/37EPS.2024.P1224

Conjugates containing homing peptide, cell penetrating peptide and a cytostatic or marker potentially useful in the diagnosis or treatment of lung cancer diseases

Piotr Jedrzejczak[a], Anna Gajda[a], Justyna Fraczyk[a], Beata Kolesinska[a]

[a]*Institute of Organic Chemistry, Łódź University of Technology, Żeromskiego 116, 90-924 Lodz, Poland*

Introduction

Lung cancer causes 23% of total cancer deaths worldwide. It is the most deadly cancer, with minimal survival. There are two known forms: non-small cell lung cancer (NSCLC) (85% of cases) and small cell lung cancer (SCLC). NSCLC is sensitive to chemotherapy; however, response varies depending on the specific therapy [1]. Oral administration of anticancer chemotherapeutic agents is often limited by first-pass metabolism. Furthermore, due to their non-targeted nature, most chemotherapeutic agents act on normal tissues, leading to adverse effects. Despite this, classical chemotherapy is used in more than 70% of patients [2]. The chemotherapy drugs used are [3]: antimetabolites (capecitabine, floxuridine, methotrexate), alkylating agents (cisplatin, carboplatin), mechlorethamine, cyclophosphamide or mitomycin C), anitimicrotubule drugs (vinca alkaloids and taxanes), topoisomerase inhibitors (etoposide, doxorubicin, mitoxantrone, novobiocin). In recent years, two groups of drugs have been introduced to the treatment of lung cancer: molecularly targeted and immunotherapy drugs. Most clinically approved therapeutic mAbs are molecularly targeted therapies, they act passively by blocking the activity of receptors or activating the immune system [4]. mAbs have limitations such as nonspecific clearance of antibodies by the reticuloendothelial system which may lead to accumulation of antibody-conjugated drugs or toxins in undesirable sites [5]. Several types of tumor-targeting ligands have been developed, including antibodies, antibody fragments, polypeptides, small molecules, and others [6]. Peptides are an attractive alternative to antibody-based targeted therapies. Unlike antibodies, peptides are easy to synthesize in large quantities [7], and their smaller size improves tissue penetration while preventing nonspecific uptake by the reticuloendothelial system. In addition, peptides can be chemically modified to alter affinity, charge, hydrophobicity, stability, and solubility, and can be optimized for in vivo use through modification. Importantly, peptides can exhibit antibody-like affinity for receptors. The aim of the research is to develop methods for the synthesis of peptide conjugates that determine selective interaction with improve transport to the cell, as well as a compound with anticancer activity.

Results and discussion

Altretamine, a derivative of 1,3,5-triazine, is an anticancer drug. The structure of altreamine was modified by using other secondary amines and by adding a peptide that penetrates cell membranes. The syntheses began with obtaining penetrating peptides using the solid-phase synthesis strategy and obtaining monosubstituted dichlorotriazine derivatives. A quaternary N-triazinylammonium salt was used as the coupling reagent [8]. Then, monosubstituted

 http://doi.org/10.17952/37EPS.2024.P1224

dichlorotriazine derivatives (substituted with a methoxy group or an appropriate amino group) were attached to the obtained peptides on the solid phase.

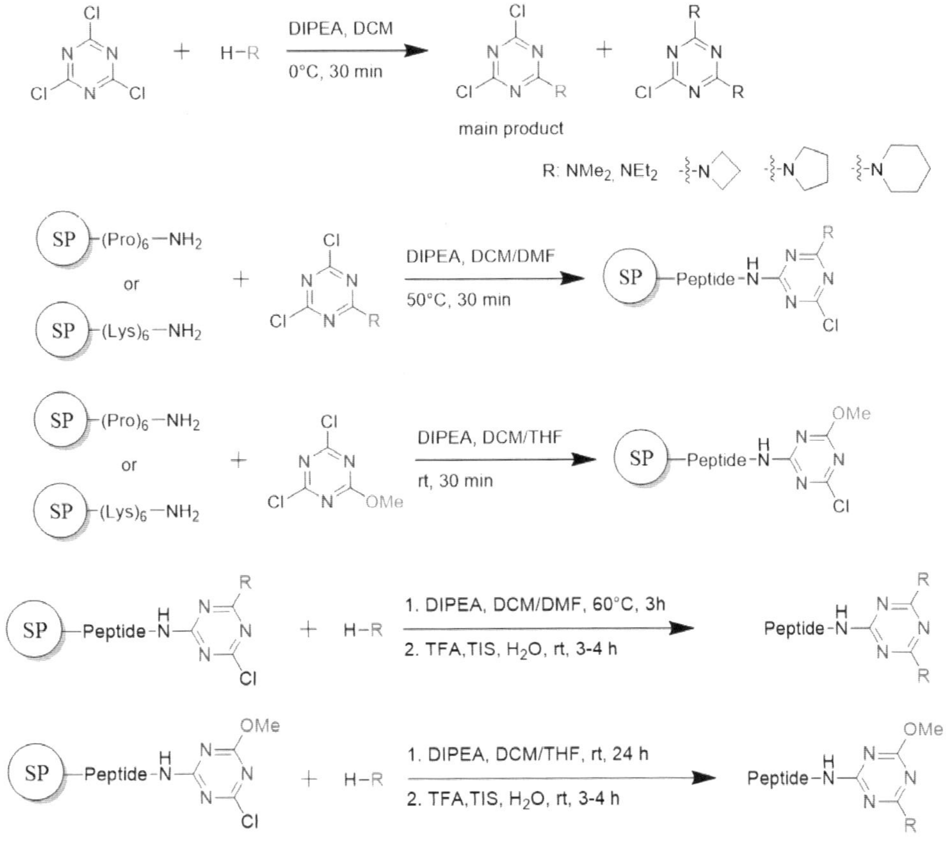

Fig. 1 Scheme for the synthesis of 1,3,5-triazine derivatives containing CPP attached to the triazine ring.

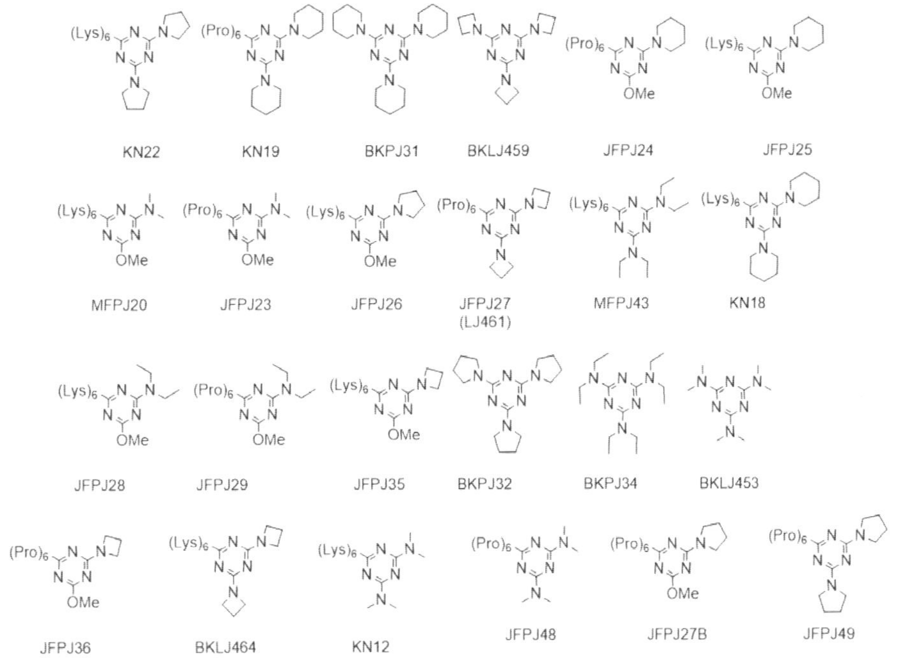

Fig. 2 Structures of the obtained 1,3,5-triazine derivatives (altretamine analogues) with CPP attached to the triazine ring.

 http://doi.org/10.17952/37EPS.2024.P1224

The final stage of the synthesis was the substitution of the last chlorine atom with an amino group and the cleavage of the compound from the solid phase. In total, 19 new compounds and aminoalkyl derivatives of 1,3,5-triazine were obtained, including altretamine as a standard.
In the final stage, the effect of the obtained compounds on A549 cells (non-small cell lung cancer, NSCLC) was assessed and the energy of interaction of the compounds with the EGF receptor was determined.

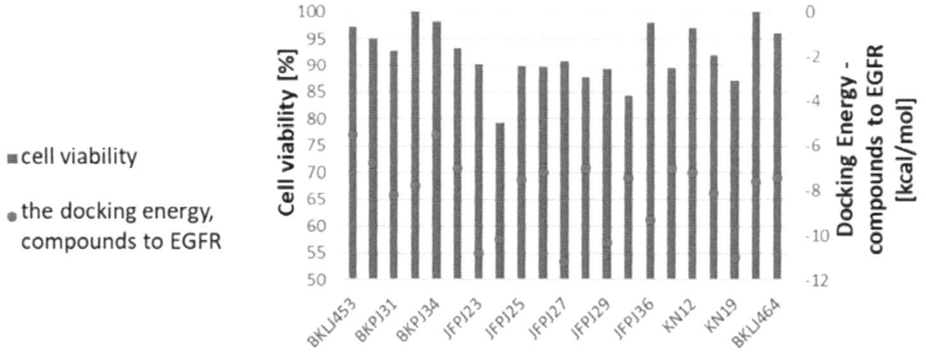

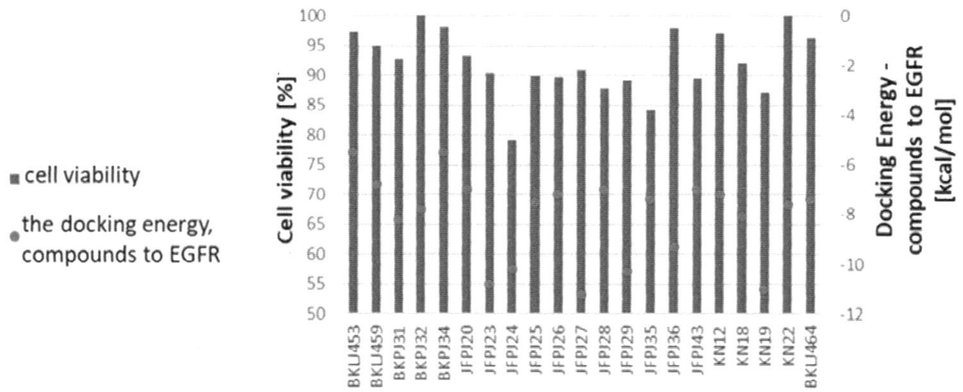

Fig. 3 The results of biological tests on the A549 cell line (NSCLC) and the calculated energy of interaction of compounds with the EGF receptor.

The biological studies carried out indicate the potential of the obtained compounds. However, due to the lack of specificity of CPP, it will be necessary to introduce an additional peptide fragment increasing the specificity towards cancer cells.

Conclusions

The results from first biological tests show increase in activity compared to control (altretamine). Further biological studies are carried out now to better characterize the obtained compounds.

Literature

[1] Jemal, A., Bray, F., Center, M.M., Ferlay, J., Ward, E., Forman, D.(2011) Global cancer statistics. *CA Cancer J. Clin.* 61, 69-90. doi: 10.3322/caac.20107.
[2] Gajarawala, S.N., Pelkowski, J.N. (2021) Telehealth Benefits and Barriers. *J. Nurse Pract.* 17, 218-221. doi.org/10.1016/j.nurpra.2020.09.013
[3] Zia, M.K., Siddiqui, T., Ali, S.S., Rehman, A.A., Ahsan, H., Khan, F.H. (2018) Chemotherapeutic Drugs and Plasma Proteins: Exploring New Dimensions. *Curr. Protein Pept. Sci.* 19, 937-947. doi.org/10.2174/1389203718666171002115547.
[4] Reichert. J.M. Valge-Archer, V.E. (2007) Development trends for monoclonal antibody cancer therapeutics. *Nat. Rev. Drug Discov.* 6, 349-56. doi: 10.1038/nrd2241.

http://doi.org/10.17952/37EPS.2024.P1224

[5] Tabrizi, M.A., Tseng, C.-M.L., Roskos, L.K. (2006) Elimination mechanisms of therapeutic monoclonal antibodies. *Drug Discov. Today* 11, 81-88. doi: 10.1016/S1359-6446(05)03638-X.
[6] Raucher, D. (2019) Tumor targeting peptides: novel therapeutic strategies in glioblastoma. Curr. Opin. Pharmacol. 47, 14-19. doi: 10.1016/j.coph.2019.01.006.
[7] Bray, B.L. (2003) Large-scale manufacture of peptide therapeutics by chemical synthesis. *Nat. Rev. Drug Discov.* 2, 587-593. doi: 10.1038/nrd1133.
[8] Kolesinska, B., Rozniakowski, K.K., Fraczyk, J., Relich, I., Papini, A.M., Kaminski, Z.J. (2015) The Effect of Counterion and Tertiary Amine on the Efficiency of N-Triazinylammonium Sulfonates in Solution and Solid-Phase Peptide Synthesis. *Eur. J. Org. Chem.* 401-408. doi: 10.1002/ejoc.201402862.

Proceedings of the 37th European Peptide Symposium
Michal Lebl, Editor
European Peptide Society, 2024
http://doi.org/10.17952/37EPS.2024.P1240

Development of Novel Angiopep-2 Conjugates for Cancer Treatment

Gábor Mező[1,2], Levente Dókus[2], Diána Vári-Mező[2,3,4], Krisztina Kiss[2], Lilla Pethő[2], Ivan Ranđelović[4], József Tóvári[4], Szilvia Bősze[2,5]

[1]Eötvös Loránd University, Institute of Chemistry, Budapest, 1117 Hungary; [2]HUN-REN–ELTE, Research Group of Peptide Chemistry, Budapest, 1117 Hungary; [3]School of Ph.D. Studies, Doctoral School of Pathological Sciences, Semmelweis University, Budapest, 1085 Hungary; [4]Department of Experimental Pharmacology and the National Tumor Biology Laboratory, National Institute of Oncology, Budapest, 1122 Hungary; [5]Department of Genetics, Cell- and Immunobiology, Faculty of Medicine, Semmelweis University, Budapest, 1089 Hungary

Introduction

Transporting therapeutics through the blood-brain barrier (BBB) is a demanding/ambitious challenge in treating brain tumors. The BBB is semipermeable; therefore, most active substances are poorly transported through this barrier, decreasing therapeutic effects. Angiopep-2 (TFFYGGSRGKRNNFKTEEY) is a peptide containing 19 amino acids that was shown to be a ligand of the low-density lipoprotein receptor-related protein-1 (LRP1). This peptide can cross the blood-brain barrier *via* receptor-mediated transcytosis and simultaneously target glioblastoma. Thereby, dual targeting may be achieved [1]. Recently, we demonstrated that substituting the three conjugation sites (two lysine side chains and the N-terminus) with drug molecules is not the best choice for developing effective peptide-drug conjugates. The substitution of Lys in position 15 decreased the cellular uptake by U87 glioma cells, while the accessibility of Lys in position 10 was somewhat hindered; therefore, the drug release was inefficient (Figure 1). However, the side chain substitution of this Lys highly increased the cellular uptake [2].

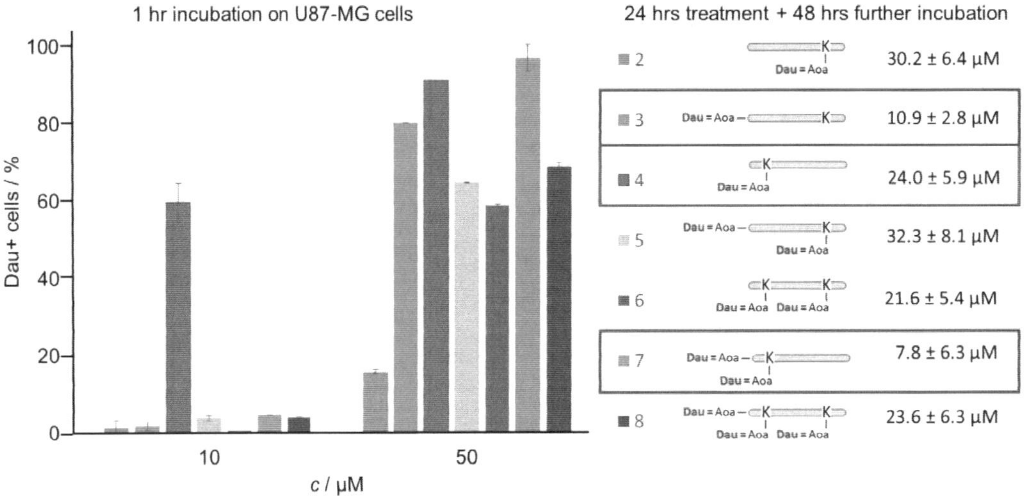

Figure 1. *In vitro* cellular uptake and cytostatic effect of Angiopep-2 – daunomycin conjugates

Thus, in this study, we developed daunomycin-Angiopep-2 conjugate in which the drug was attached to the N-terminus *via* oxime-linkage, and the side chain of Lys in position 10 was substituted with butyric acid (Dau=Aoa-TFFYGGSRGK(Bu)RNNFKTEEY-OH). The modification was used earlier efficiently in order to increase the cellular uptake and the antitumor effect of peptide-drug conjugates [3].

Results and Discussion

The Angiopep-2 peptide was built up on Wang resin by standard SPPS with Fmoc/tBu strategy. At the end of the sequence isopropylidene protected aminooxyacetic acid (Aoa) was attached similarly with Oxima Pure/HOBt coupling reagents. After the peptide cleavage from the resin the isopropylidene group was removed with 1M methoxyamine in 0.2M NH_4OAc buffer at pH 5 followed by purification and ligation with daunomycin (Dau) under the same condition. In the case of the butyrylated derivative the side chain of Lys in position 10 was protected with Mtt group that was removed prior the final cleavage with 2% TFA/DCM solution followed by butyrylation using butyric anhydride. The further reaction steps were the same as in the case of non-butyrylated derivative.

 http://doi.org/10.17952/37EPS.2024.P1240

The cellular uptake and the *in vitro* cytostatic effect of butyrylated conjugate was significantly higher (IC_{50} value was 0.678 ± 0.089 μM) than the non-butyrylated version (1.418 ± 0.153 μM), which confirmed our hypothesis (Figure 2).

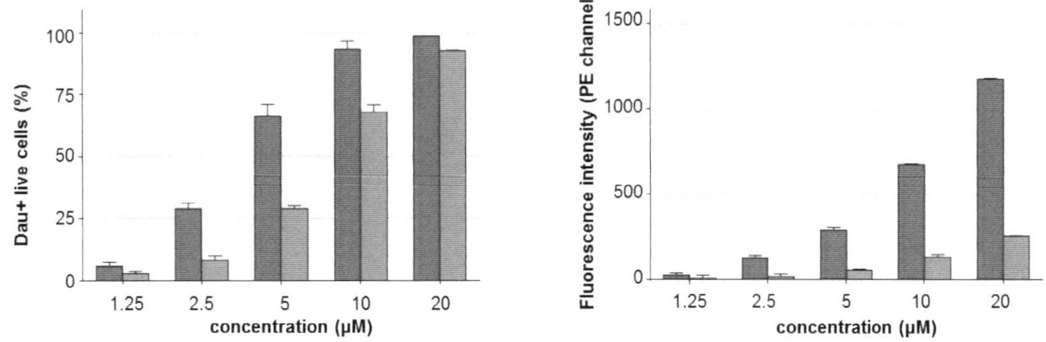

Fig. 2. Cellular uptake of Angiopep-2 – daunomycin conjugates (red: butyrylated; orange: non-butyrylated)

The short-chain acidic acid-containing conjugate showed a bit higher but insignificant (only 2%) passage through the BBB model [4] without LRP1. The conjugates were not cytotoxic on the endothelial cells.

According to these results, the conjugates were also investigated *in vivo* on U87MG *s.c.* tumor model. Dau had no tumor growth inhibition effect in the MTD concentration (1mg/kg) on this type of tumor model. However, the conjugates showed slight but not significant inhibition (10-15%) using a dose of 10 mg Dau content of conjugate/kg. The butyrylated conjugates elevated the tumor doubling time (11%) a bit and showed lower toxicity than the free Dau (Figure 3). Therefore, the butyrylated Angiopep-2 - drug conjugates might be good candidates for targeted tumor therapy, but more potent drug molecules should be used.

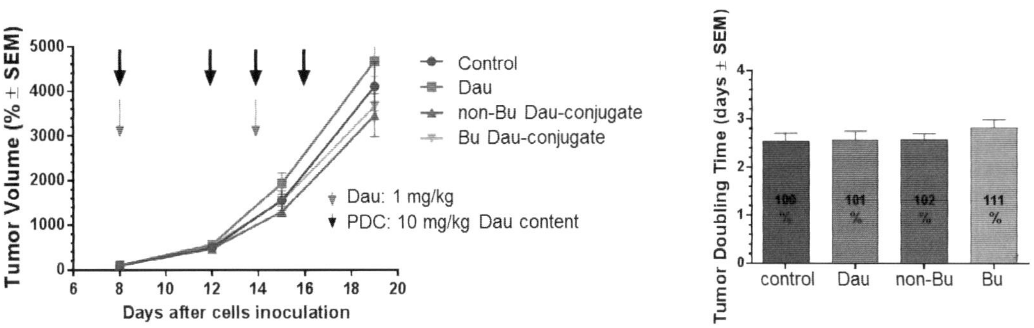

Fig 3. Tumor growth inhibition (black arrows: treatments with conjugates, red arrows: treatments with Dau) and tumor doubling time

Acknowledgments

The research was supported by the National Research, Development and Innovation Office under grants NKFIH K146039 and the National Laboratories Excellence program, as part of the National Tumor Biology Laboratory project (2022-2.1.1-NL-2022-00010).

References

1. Demeule, M., et al. *J. Pharmacol. Exp. Ther.* **324**, 1064-1072 (2008), doi: 10.1124/jpet.107.131318
2. Pethő, L., et al. *Int. J. Mol. Sci.* **24**, 3106 (2023), doi: 10.3390/ijms24043106.
3. Hegedüs, R., et al. *Eur. J. Med. Chem.* **56**, 155-165 (2012), doi: 10.1016/j.ejmech.2012.08.014
4. Baranyai, Zs., et al. *J. Med. Chem.* **64**, 2982-3005 (2021), doi: 10.1021/acs.jmedchem.0c01399

https://doi.org/10.17952/37EPS.2024.P1272

Application of the MALDI-MS and MALDI-IMS in the identification of the GPR146 receptor in tissue preparations

Joanna Waśko[1], Aleksandra Ryk[2], Wojciech Fendler[2], Beata Kolesińska[1]
[1]Institute of Organic Chemistry, Lodz University of Technology, Zeromskiego 116, 90-543 Lodz, Poland,
[2]Department of Biostatistics and Translational Medicine, Medical University of Lodz, Mazowiecka 15, 92-215 Lodz, Poland

Introduction

G protein-coupled receptor 146 (GPR146, PGR8) belongs to membrane protein with molecular mass around 36.6 kDa. Its activity is regulated by seven typical transmembrane domains (Fig. 1). GPR146 is highly expressed in the livers (especially in hepatocytes) and adipose tissues of humans and mice. The activation of a G-protein coupled receptor can trigger a number of intracellular signalling cascades and responses. Which pathway is activated varies depending on the class of the coupled protein. Regardless of class, the G proteins initiate interactions and relocation of cellular components.

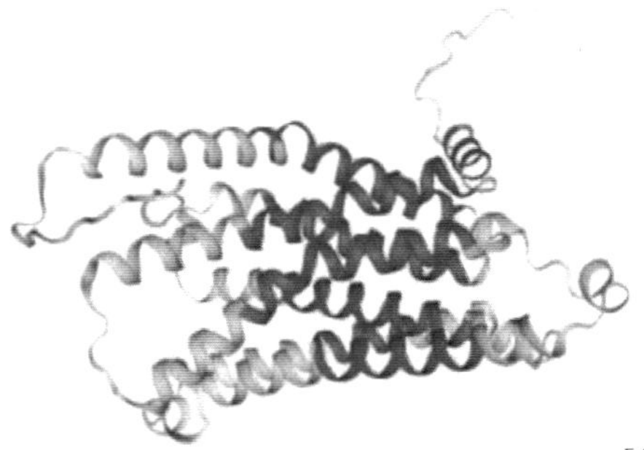

Fig.1. The structure of GPR146 receptor. [5]

Despite the fact that signalling mechanisms have been well studied and established, it has been increasingly clear that GPR146 signalling biology is remarkably more complex than originally thought. Moreover, the number of potential ligands, compatible to GPR146 receptor could be also more, which makes it possible to expand the use of this protein in the diagnosis and treatment of many diseases. Based on human genetic findings and on mouse studies, the G protein-coupled receptor emerges as a promising target against hypercholesterolemia and atherosclerosis [1,2]. It is also known their potential in diabetes treatment – current GPR-targeting therapies include glucagon-like peptide 1 (GLP-1) based agonists, the related DPP4 antagonists, bromocriptine, and serotonin agonists [3]. In turn, proinsulin C-peptide has gained recognition as a potential therapeutic target for the treatment of metabolic disorders and there are some reports indicate on their correlation with GPR146 receptor activity [1,2,4]. Therefore, in order to develop more effective methods of treating type I and II diabetes, it is important to know the correlation between receptor expression and the stage of disease advancement.

Results and Discussion

The attempts were made to detect the transmembrane receptor GPR146 in biological material (fresh tissue and tissue homogenates) collected from various organs of patients with known insulin resistance using matrix-assisted laser desorption/ionization spectrometry (MALDI-MS) and imaging mass spectrometry (MALDI-IMS) equipped in time of flight analyzer (TOF). Sinapic acid (SA) was used as a matrix in both techniques, and a mixture of water and acetonitrile with addition of 1% trifluoroacetic acid as the solvent (in a volume ratio of 1:1).

In the first stage, MALDI-MS measurements were performed for tissue homogenates, collected from patients with known insulin resistance, from: the saphenous vein (Z), right atrium of the heart (U), para-aortic fat tissue (TKO), epicardial fat tissue (TKN) and postthymic fat tissue

(TKP). Furthermore, MS spectra for the standard sample of GPR146 were also recorded and compared with homogenates results (Fig.2).

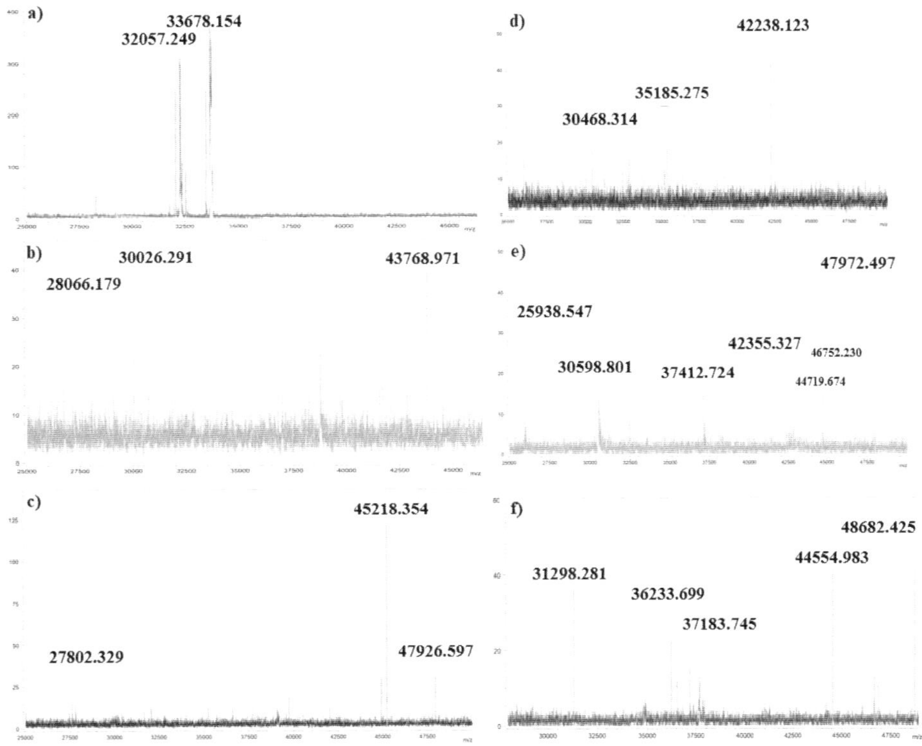

Fig.2. The MALDI-MS spectra registered for the following homogenates: a) TKN, b) TKO, c) Z; d) TKP; e) U and for f) GPR146 standard sample. SA concentration 20 mg/ml.

The presence of GPR146 depends on the type of homogenate – for that receptor the most characteristic and repetitive peaks in MALDI-MS analyses were detected in postthymic fat tissue (TKP) and right atrium of the heart (U). On the other hand the results registered for the saphenous vein (Z) didn't confirm expression of the receptor within tissue homogenate. Nextly, MALDI-IMS imaging was performed for tissue specimens taken from a healthy patient (reference sample) and from a patient with diagnosed metabolic disorders (Fig.3). It was compared occurrence and intensity of peaks characteristic for GPR146 within both of them. Firstly, the ionization efficiency was higher in the case of healthy tissue – probably, changes occurring in the diabetic's body definitely

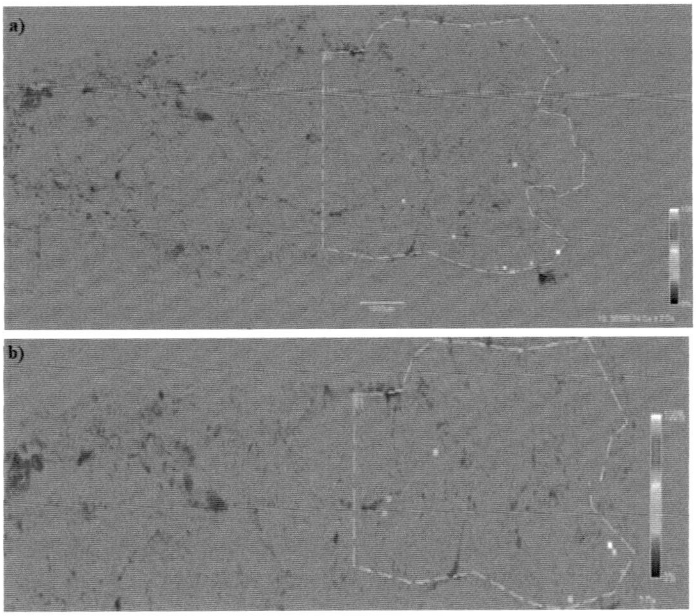

Fig.3. GPR146 distribution within: a) diabetic tissue, b) healthy tissue.

modify its structure. In addition, the occurrence of GPR146 was more intensive and visible in the pancreas section taken from a patient with diagnosed insulin resistance.

Conclusions

Based on the presented results we concluded that GPR146 expression within human tissues may be higher during metabolic disorders. Moreover the same tendency we observed for peptide C in previous studies, thus both of them should be take into account during designing innovative and more effective methods for treatment of diabetes mellitus. In samples collected from circulatory system the largest number of peaks corresponding to transmembrane protein were identified, which confirms the relationship between diabetes and heart diseases. Moreover, increased expression of the receptor in tissue taken from a patient with known insulin resistance may also indicate possible correlations between the amount of C-peptide and the GPR146 receptor.

Literature

[1] Lindfors, L., Sundström, L., Fröderberg Roth, L., Meuller, J., Andersson, S., Kihlberg, J. (2020). Is GPR146 really the receptor for proinsulin C-peptide? Bioorganic and Medicinal Chemistry Letters, 30, 127208. doi: 10.1016/j.bmcl.2020.127208

[2] Kolar, G.R., Grote, S.M., Yosten, G.L.C. (2017). Targeting orphan G protein coupled receptors for the treatment of diabetes and its complications: C-peptide and GPR146. J Intern Med. 281(1), 25–40. doi:10.1111/joim.12528.

[3] Ahren, B. (2009). Islet G protein-coupled receptors as potential targets for treatment of type 2 diabetes. Nat Rev Drug Discov. 8(5), 369–385.

[4] Yosten, G.L.C., Kolar, G.R. (2015). The Physiology of Proinsulin C-Peptide: Unanswered Questions and a Proposed Model. PHYSIOLOGY, 30, 327–332. doi:10.1152/physiol.00008.2015

[5] https://www.proteinatlas.org/ENSG00000164849-GPR146/structure (accsess date: 02.07.2024)

Proceedings of the 37th European Peptide Symposium
Michal Lebl, Editor
European Peptide Society, 2024 *http://doi.org/10.17952/37EPS.2024.P1273*

Searching for biomarkers characteristic for severe stage of COVID-19 disease

Aleksandra Czerchawy[1], Joanna Waśko[1], Beata Kolesińska[1]

[1] *Institute of Organic Chemistry, Faculty of Chemistry, Lodz University of Technology, Zeromskiego 116, Lodz, 90-543, Poland*

Introduction

Biomarkers, also known as biological markers, are defined as objectively measured and evaluated molecules (e.g. amino acids, peptides, proteins, enzymes, lipids, sugars, antibodies, hormones etc.) that act as indicators of typical biological processes, pathogenic processes or pharmacological responses. Their estimation can provide a valuable source of information about a patient's health condition, thus allowing for better diagnosis, prognosis and treatment options. Biomarkers analysis in physiological fluids (e.g. blood, urine, saliva, sweat, tears, faeces or exhaled air) can also offer a quick, painless, non-invasive and cost-effective method for disease detection and monitoring. However, this approach so far has many limitations due to the complexity and wide dynamic range of compounds present in bodily fluids. A promising alternative could be the direct identification of characteristic compounds from tissue sections, which would provide a comprehensive molecular view in a histology-related context [1].

Proteomics as a large-scale study of interactions, function, composition and structures of proteins could offer extensive insights into the molecular and cellular mechanisms of diseases. Nevertheless, the diversity of protein sequences and post-translational modifications makes it a challenging discipline. Fortunately, impressive technological advancements combined with bioinformatics have significantly progressed proteomics, aiding in compound characterization. The most commonly applied methods contain electrophoresis and high-throughput technologies such as liquid chromatography with mass spectrometry (tandem MS, MALDI-TOF) [2].

Under these circumstances, MALDI-MSI has gained particular attention due to its many advantages. This label-free technique generates two-dimensional ion density maps, enabling a spatial distribution of various biomolecules, including pharmaceuticals, lipids, peptides and proteins, through a tissue section in a single run. It allows for the simultaneous screening of hundreds of biological compounds, even low molecular weight compounds fig(e.g. drugs, metabolites) with a high sensitivity for multiple samples placed in a single target plate What's more, an easy sample preparation and a soft-ionization process simplifies data collection and interpretation [3].

The COVID-19 pandemic highlighted the need for accurate diagnostic tools. Utilizing modern techniques in biomarker search and recognition would help indicate the disease, assess the infection severity, and identify at-risk patients. The collected information could help in making treatment decisions and determining admission or intensive care criteria [4].

Results and Discussion

The aim of the studies conducted in the Institute of Organic Chemistry was the search and identification of potential protein and peptide biomarkers indicating the SARS-CoV-2 virus infection. The research focused on analysing the lung tissues that could indicate characteristic compounds for the COVID-19 disease, the advancement stage or other complications related to the disease course (e.g. lung fibrosis) and involved preparing lung tissue samples taken from a person who died of severe SARS-CoV-2 virus infection for MALDI-MSI imaging studies. Firstly, the samples were covered with two matrices, α-cyano-4-hydroxycinnamic acid (CHCA) and 2,5-dihydroxybenzoic acid (DHB) to detect compounds with low and high molecular masses, respectively. Matrix-coated samples were next subjected to MALDI-MSI analysis [5]. For some values revealed in a range of molecular ions, MALDI-MSI imaging was performed to show compound distribution within the tissue sample (Figure 1). Results combined with bioinformatic methods allowed for the identification of protein fragments involved in the process of replication and degradation of the SARS-CoV-2 genome, as well as other peptides associated with inflammatory processes, or even the fragment of hACE2 enzyme, which serves as a receptor for coronavirus particles (Table 1).

Fig. 1. *Picture showing the distribution of a ~1199 Da peptide identified in lung tissue using MALDI-MSI*

Table 1. *Representation of peptides identified during MALDI-MSI analysis as potential biomarkers for COVID-19 disease.*

	Protein	Fragment	m/z	Commentary
Inflammation	Vitamin K-dependent protein C (PROC)	HSSLER	729,29	May be involved in the pathogenesis of COVID-19 (lung damage and thrombogenicity). Involved in the coagulation process and the metabolism of elastic fibres
	Pulmonary surfactant-associated protein A1 (SFTPA1)	NCLYSR	755,37	Binding surfactant phospholipids lowering the surface tension in alveoli. Associated with lung fibrosis, a side effect of SARS-CoV-2 infection
	Fibrinogen alpha chain (FGA)	DSGEGDFL AEGGGVR	~1464	Increased levels observed for COVID-19 cases correlating with excessive inflammation and disease severity [6]
	Serum amyloid A-1/ A-2 protein (SAA1/ SAA2)	FFSFLGEAF DGAR	~1464	Increased levels of amyloid A observed in human serum of patients with severe SARS-CoV-2 virus infection [7]
	Complement component 6 (C6)	ALNHLPLEY NSALYSR		Constituent of the membrane attack complex playing a role in the innate and adaptive immune response
	Apolipoprotein A-II (APOA2)	VKSPELQAE AK	~1199	Decreased level of APOA1, APOA2 and APOC2 in COVID-19 plasma patients [8]
	Apolipoprotein A-I (APOA1)	DSGRDYVS QFEGSALG K	~1816	
	Apolipoprotein C-II (APOC2)	ESLSSYWES AK	~1286	
hACE2	Angiotensin-converting enzyme type 2 (ACE2)	LFNMLR; HEGPLHK	793,44 817,50	Receptor interacting with RBD of the S1 protein subunit of SARS-CoV-2
Replication and Replication and degradation of the viral genome	Replicase polyprotein 1ab (fragment)	NLYDK	651,64	Involved in the transcription and replication of viral RNAs
	Cystic fibrosis transmembrane conductance regulator (CFTR)	WFQMR	769,35	CFTR expression/function is involved in the regulation of SARS-CoV-2 replication
	E3 ubiquitin-protein ligase HUWE1 (HUWE1)	SMLNFLK; HLGYTLCK	853,78 935,74	Protein inducing ubiquitination and degradation of the MERS-CoV protein ORF3 (important for SARS-CoV-2 infection)

Conclusions

Progress in biomarker recognition and proteomic technologies shows great promise for a better understanding of disease pathologies and improving diagnostic procedures, particularly in the context of COVID-19. Nonetheless, it is still a developing field of study and involves significant costs and long-term examination. The analysis of samples is highly dependent on the sample collection and preparation, and the longevity of probes is also a consideration. Result interpretation is not an easy task and requires in-depth knowledge of multiple inflammatory processes. Currently, analyses are so far compared with standard approaches, but ongoing advancements may simplify the process.

Acknowledgements

The research was founded by "Young Scientist Fund" no. W-3D/FMN/16G/2023.
Samples of lung tissues and respiratory filters were taken from the Military Institute of Medicine in Warsaw and the Infectious Diseases Hospital in Lodz.

References

[1] Minerva, L., Ceulemans, A., Baggerman, G., Arckens, L. *Proteomics Clin. Appl.* **6**, 581-595 (2012), https://doi.org/10.1002/prca.201200033

[2] Al-Amrani S., Al-Jabri, Z., Al-Zaabi, A., Alshekaili, J., Al-Khabori, M. *World J Biol Chem.* **12(5)**, 57-69 (2021), https://doi.org/10.4331/wjbc.v12.i5.57

[3] Zhu, X., Xu, T., Peng, C. *Front. Chem.* **9**, 782432 (2022), https://doi.org/10.3389/fchem.2021.782432

[4] Samprathi, M., Jayashree, M. *Front. Pediatr.* **8**, 7607647 (2021), https://doi.org/10.3389/fped.2020.607647

[5] Aichler, M., Walch, A. *Lab. Invest.* **95**, 422-431 (2015), https://doi.org/10.1038/labinvest.2014.156
[6] Sui, J., Noubouossie, D.F., Gandotra, S., Cao, L. *Front. Cell. Infect. Microbiol.* **11**, 734005 (2021), https://doi.org/10.3389/fcimb.2021.734005
[7] Jana, A.K., Greenwood, A.B., Hansmann, U.H.E. *J. Phys. Chem. B.* **125 (32)**, 9155–9167 (2021), https://doi.org/10.1021/acs.jpcb.1c04871
[8] Faguer, S., Del Bello, A., Danet C., Renaudineau Y., Izotep J., Kamar, N. *Front. Pharmacol.* **13,** 936659 (2022), https://doi.org/10.3389/fphar.2022.936659

https://doi.org/10.17952/37EPS.2024.P1292

Synthesis and assessment of amyloidogenic activity of the C-peptide fragments

Joanna Waśko[1], Beata Kolesińska[1]
[1]Institute of Organic Chemistry, Faculty of Chemistry, Lodz University of Technology, Zeromskiego 116, 90-543 Lodz, Poland

Introduction

In the recent years, the research on amyloidogenic peptides have become a key area of scientific research in the field of biotechnology and medicine. The amyloidogenic peptides have significantly contribution in understanding the pathological mechanisms associated with neurodegenerative disorders such as Alzheimer's and Parkinson's disease. Moreover, it has been found that undesirable aggregation of metabolic hormones (insulin and amylin) could be also a source of the progress of type I and II diabetes. It is known that the diabetes mellitus is a group of physiological dysfunctions characterized by hyperglycaemia resulting directly from insulin resistance (in the case of type 2 diabetes mellitus-T2DM), inadequate insulin secretion/production, or excessive glucagon secretion (in type 1 diabetes mellitus-T1DM) [1,2]. Although the aetiology of diabetes may differ from T1DM to T2DM, common features, namely formation of insoluble amyloid deposits within different tissue, may occur during the progression of the both diseases. It has been proven that amyloid deposits consist largely of aggregated insulin molecules (especially during long-term subcutaneous application of the hormone during T1DM) and both insulin and amylin chains in the case of T2DM. Amylin is co-secreted with insulin from pancreatic β cells, and due to its tendency to aggregate within pancreatic islet cells in type 2 diabetes mellitus (T2DM) is so-called as human amyloid polypeptide (hIAPP) [3,4]. The amyloidogenic properties of peptides could be also favoured by factors like acidic pH, drastic changes in temperature, contact with hydrophobic surfaces or variations in ionic strength of solution. Although there are reports which put hypothesis that amyloid deposits may also include other structures. Currently, researchers are increasingly focusing on the tendency to aggregate C-peptide and its input on another metabolic disorders [5,6].

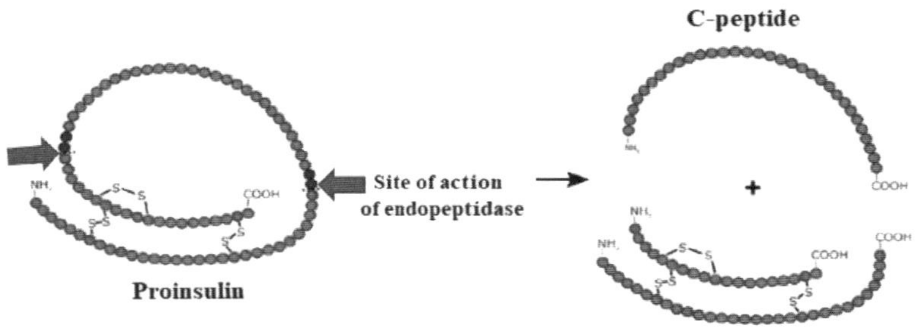

Fig.1. Synthesis of insulin and C-peptide at the cellular level.

Initially, peptide C, produced in the cell during the proteolytic transformation of proinsulin into insulin and co-secreted in equimolar amounts into the circulatory system together with insulin, was considered as a by-product whose action did not significantly affect the functioning of the body. However, its activity has now been confirmed at the stage: a) promoting proper folding of proinsulin; b) initiating the process of creating disulphide bridges within the hormone; c) supporting interactions enabling controlled secretion of insulin from the pancreas. On the other hand, there is some controversy regarding reported effects of the C-peptide. Its beneficial effects have been demonstrated in long-term treatment in type 1 diabetes. In contrast to this, C-peptide in type 2 diabetes could induce proinflammatory and proatherogenic effects.

Results and Discussion

Based on these inconsistencies, one of the directions of our research includes studies on the tendency to aggregation of C-peptide fragments under *in vitro* conditions.

To verify this hypothesis, C-peptide was divided into five shorter fragments (Table 1). As it was identified for insulin [7], we also would like to detect the shortest potential sequences within C-peptide responsible for the growth of amyloid fibril (so-called hot-spot fragments).

Table 1. The synthesized C-peptide fragments.

(1)	H_2N-[57]EAEDLQV[63]-OH
(2)	H_2N-[64]GQVELG[69]-OH
(3)	H_2N-[70]GGPGAG[75]-OH
(4)	H_2N-[76]SLQPLA[81]-OH
(5)	H_2N-[82]LEGSLQ[87]-OH

The proposed peptide chains were synthesized according to the Fmoc/tBu methodology on a 2-chlorotrityl chloride resin using a triazine coupling reagent [8]. In the second stage of the research, a spectrophotometric and microscopic tests were carried out to assess their tendency to aggregation.

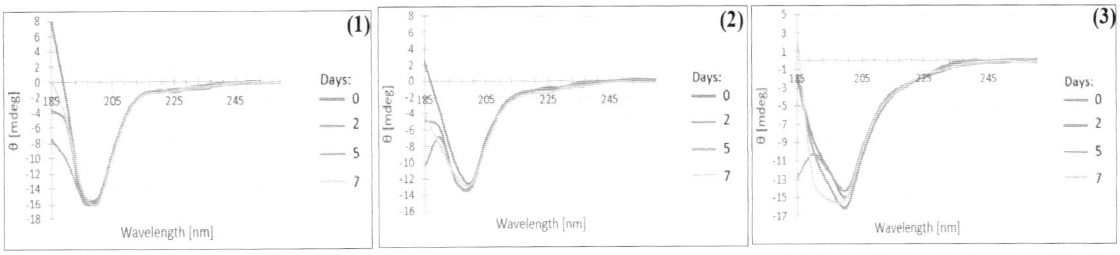

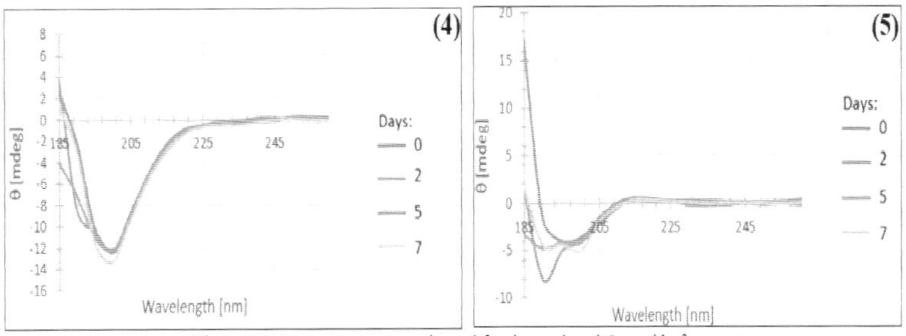

Figure 1. CD measurements registered for the analyzed C-peptide frgaments.

Firstly, circular dichroism measurements (CD) were performed for the peptides (1-5). The samples were dissolved in phosphate buffer (pH 7.2) and then incubated at 37.4°C. Based on the registered CD spectra we conclude that first four peptides (1-4) showed stable secondary structure during 7 days. In turn, the fragment (5) was the most susceptible to conformational changes. For the

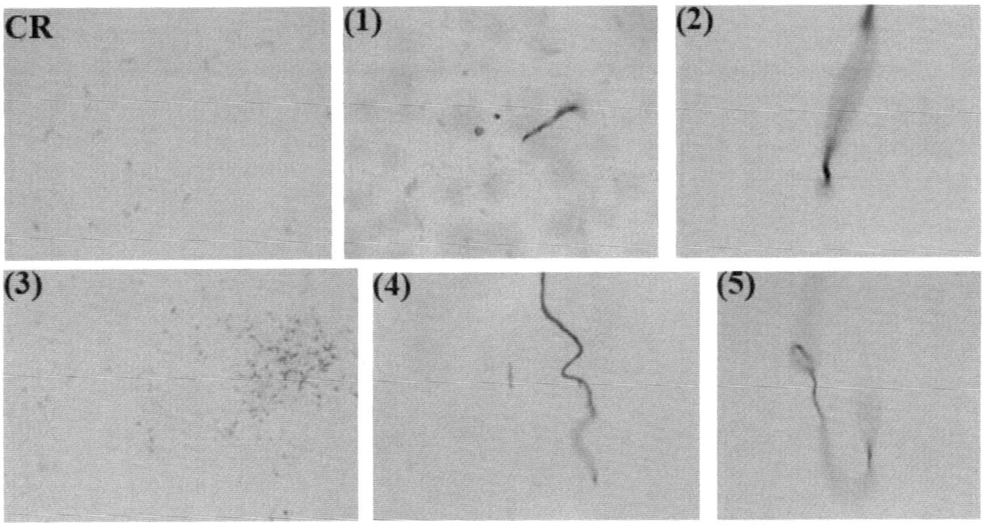

Figure 2. The microscopic analysis for samples incubated with CR.

microscopic studies carried out for the incubated samples, additionally stained with a dye that selectively interacts with amyloid fibers (namely Congo red, CR), the highest density of fibrillar structures was also identified in the fifth, C-terminal sequence, marked in locants 82-87. This is most likely related to the presence of the active site in this area, information about which can be found in the literature [9].

Moreover, the unmature amyloid fibers were observed for the samples (1), (2) and (4) but there were unstable. On the other hand, in the case of the third peptide (70-75), the predisposition to form aggregates was the weakest compared to the rest of the synthesized amino acid chains.

Conclusions

The research carried out so far point out the tendency to aggregate some of the analyzed sequences, while opening a new research approach to searching for new functions of C-peptide as a biomarker in the diagnosis of diseases related to metabolic disorders. As part of further experiments, it is planned to repeat the aggregation tests for structures composed of at least seven amino acids and at a reduced pH.

Literature

[1] Wahren, J., Ekberg, K., & Jörnvall, H. (2007). C-peptide is a bioactive peptide. Diabetologia, 50, 503-509. https://doi.org/10.1007/s00125-006-0559-y

[2] Krause, M., De Vito, G. (2023). Type 1 and type 2 diabetes mellitus: commonalities, differences and the importance of exercise and nutrition. Nutrients, 15, 4279. https://doi.org/10.3390/nu15194279

[3] Alrouji, M., Al-Kuraishy H.M., Al-Gareeb, A.I., Alexiou, A., Papadakis, M., Saad, H.M., El-Saber Batiha, G. (2023). The potential role of human islet amyloid polypeptide in type 2 diabetes mellitus and Alzheimer's diseases. Diabetology & Metabolic Syndrome, 15 (101), 1-16. https://doi.org/10.1186/s13098-023-01082-1

[4] Das, A., Shah, M., Saraogi, I. (2022). Molecular aspects of insulin aggregation and various therapeutic interventions. ACS Bio Med Chem Au, 2, 205–221. https://doi.org/10.1021/acsbiomedchemau.1c00054

[5] Vasic, D., Walcher, D. (2012). C-Peptide: A new mediator of atherosclerosis in diabetes. Mediators of Inflammation, 2012, 1-8. https://doi:10.1155/2012/858692

[6] Tsiolaki, P.L., Louros, N.N., Zompra, A.A., Hamodrakas, S.J., Iconomidou, V.A. (2016). Unraveling the aggregation propensity of human insulin C-Peptide. PeptideScience, 108 (2), 1-8. https://doi.org/10.1002/bip.22882

[7] Swiontek, M., Fraczyk, J., Wasko, J., Chaberska, A., Pietrzak, L., Kaminski Z.J., Szymanski, L., Wiak, S., Kolesinska, B. (2019). Search for new aggregable fragments of human insulin. Molecules, 24, 1600, 1-20. https://doi.org/10.3390/molecules24081600

[8] Kolesińska B., Rożniakowski K.K., Frączyk J., Relich I., Papini A.M., Kamiński Z.J. (2015). The Effect of Counterion and Tertiary Amine on the Efficiency of N-Triazinylammonium Sulfonates in Solution and Solid-Phase Peptide Synthesis. Eur. J. Org. Chem., 401-408. doi.org/10.1002/ejoc.201402862

[9] Munte, C. E., Vilela, L., Kalbitzer, H. R., & Garratt, R. C. (2005). Solution structure of human proinsulin C-peptide. The FEBS Journal, 272 (16), 4284-4293. https://doi.org/10.1111/j.1742-4658.2005.04843

https://doi.org/10.17952/37EPS.2024.P1306

Tailoring the self-assembly of cyclic peptide scaffolds for prostate cancer imaging

Fátima Santillán,[1] Carlie Charron,[2] Betty C. Galarreta,[3] Leonard G. Luyt[4]

[1] Department of Chemistry, University of Western Ontario, 1151 Richmond St, London, Ontario N6A 5B7, Canada
[2] Department of Chemistry, Dalhousie University, 6274 Coburg Road, Halifax, Nova Scotia B3H 4R2, Canada
[3] Departamento Académico de Ciencias - Sección Química, Pontificia Universidad Católica del Perú, Av. Universitaria 1801, San Miguel, Lima 15088, Perú
[4] Departments of Chemistry and Departments of Medical Imaging and Oncology, University of Western Ontario, University of Western Ontario, Centre; 1151 Richmond St, London, ON, N6A 3K7, Canada. Verspeeten Family Cancer Centre, London Health Sciences, 800 Commissioners Rd. E., London, ON, N6A 5W9, Canada.

Introduction

Advancements in the design of peptide self-assemblies have opened the potential for nanostructuresto be used in cancer diagnosis and imaging. The use of peptide nanomaterials as imaging probes represents a promising alternative owing to its tailorable design that provides specificity towards molecular targets.[1-4] In the context of cancer diagnostics, optical imaging holds immense potential, where precise targeting of biomarkers, like the prostate-specific membrane antigen (PSMA), is crucial for cancer detection.

In our exploration of optical imaging agents for cancer, we focused on the design and co-assembly of cyclic peptide templates that generate nanotubes. These nanotubes were fabricated through the pH-triggered self-assembly of D/L-alternating cyclic octapeptides.[1] The main sequence is derived from cyclo-(D-Leu-Lys-D-Leu-Tyr)$_2$. The surface of the peptide was modified to enhance its functionality, leading to three additional peptides that were critical for building the nanoprobe. The first modification involves the attachment of polymeric chains (m-PEG) to reduce nanotube aggregation. The second peptide featured the conjugation of a fluorescent dye as a signaling component. Lastly, our targeting peptide incorporated a modified PSMA-targeting ligand for cancer imaging. All the peptides were purified by reverse-phase high-performance liquid chromatography (LC-MS) and analyzed using high-resolution mass spectrometry.

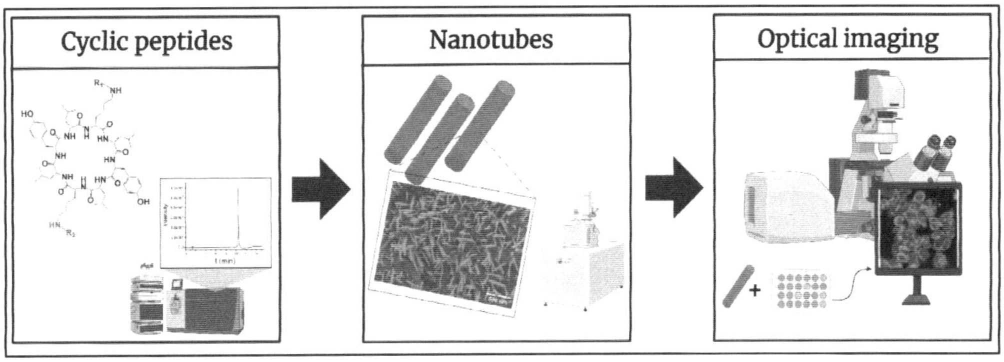

Fig. 1 General schematic representation of the research conducted

Results and Discussion

The self-assembly behavior of all the cyclic peptides and the co-assembled systems was assessed using field emission scanning electron microscopy (FESEM). FESEM imaging revealed diameters ranging from 13 to 30 nm and lengths of 100-300 nm for most peptide systems. Notably, our targeting nanotubes, formed through the co-assembly of all four peptides, exhibited an average diameter of 28 nm and length values of 100-400 nm (Figure 2b). Circular dichroism spectroscopy revealed the presence of β-sheet-like assemblies, which were stable under room and physiological temperature conditions (Figure 2c).

Confocal fluorescence microscopy evaluated the targeting capabilities of the co-assembled cyclic peptides nanotubes (Figure 2d). To this end, two cancer cell lines, PC3 (PSMA-low) and LNCaP (PSMA-high) were selected. Confocal fluorescence images of LNCaP cells, treated with our targeting nanotubes, exhibited a significantly heightened fluorescence signal on cell membranes compared to non-targeting nanotubes. This confirms the interaction of our developed nanoprobes with PSMA overexpressed in LNCaP cells. Further analysis via Z-stacking indicated partial internalization of the targeting nanotubes, suggesting a potential endocytic mechanism. Confocal fluorescence images of PC3 cells, treated with our targeting nanotubes, exhibited low fluorescence signal as was expected due to the reduced presence of PSMA in this cell line.

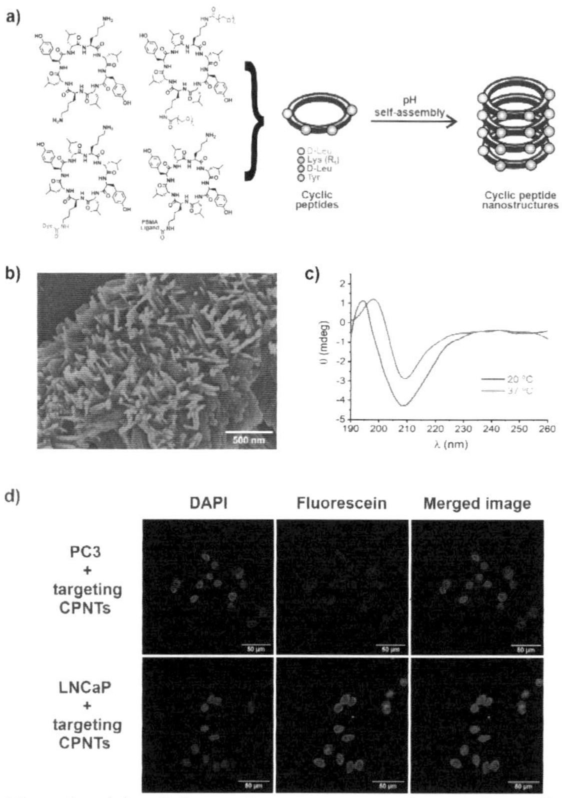

Fig. 2 (a) Schematic representation of self-assembly. (b) FESEM images of targeting nanotubes. (c) CD spectra at room and physiological temperatures. (d) Confocal fluorescence microscopy images of PC3 and LNCaP cells with targeting nanotubes after incubation for 2 h at 37 °C.

In conclusion, our findings underscore the enhanced target specificity that can be achieved by engineering peptide nanotubes with tailorable functionalities, expanding the use of co-assembled peptide nanoprobes to the realm of molecular imaging and offering possible pathways towards clinical translation.

Acknowledgments

This research was supported by the Natural Sciences and Engineering Research Council of Canada (NSERC) and the Collaborative Program in Molecular Imaging (Western University). The authors would like to thanks Surface Science Western for the training and access to field emission scanning electron microscopy.

References

1. Insua, I., Montenegro, J. *J. Am. Chem. Soc.* **142**, 300-307 (2020). https://doi.org/10.1021/jacs.9b10582

2. A. Méndez-Ardoy, A. Bayón-Fernández, Z. Yu, C. Abell, J. R. Granja, J. Montenegro, *Angewandte Chemie* **132,** 6969 (2020). https://doi.org/10.1002/anie202000103

3. S. Díaz, I. Insua, G. Bhak, J. Montenegro, *Chem. Eur. J.* **26,** 14765 (2020). https://doi.org/10.1002/chem.202003265

4. Panigrahi, B., Singh, R.K., Suryakant, U., Mishra, S., Potnis, A.A., Jena, A.B., Kerry, R.G., Rajaram, H., Ghosh, S.K., Mandal, D. *Eur. J. Pharm. Sci.* **171**, 106125 (2022). https://doi.org/10.1016/j.ejps.2022.106125

https://doi.org/10.17952/37EPS.2024.P2046

Divalent metal ions-directed release of *C*-terminally modified peptides at 60 °C under mild conditions

Rebeca G. Moura, Thiago de Souza Freire, M. Terêsa Machini

Department of Biochemistry, Institute of Chemistry, University of São Paulo, Av. Prof. Lineu Prestes, 748, 05508-900, São Paulo, Brazil
rebeca.garcia.moura@usp.br; mtmachini@iq.usp.br

Introduction

C-terminally modified peptides (*C*-TMPs) are of great interest in the field of therapeutic peptides as modification of such macromolecules can favorably impact its potency, chemical stability, hydrophobicity and biological barriers permeability.[1] The synthesis of *C*-TPMs can be complex and time-consuming. In fact, some steps of the synthetic routes employ acidic or basic conditions, leading to *C*-terminal amino acid enantiomerization[2a,b] and other side-reactions.

In the 1990s, Machini-Miranda et al. first observed transesterification of the *N*-acylated *C*-esterified tripeptide Moz-Asn-Leu-Gly-OEt in MeOH/H_2O containing calcium salts. These authors also proved the applicability such Ca^{2+}-promoted transesterification by using this new procedure to produce various di- and tripeptide esters.[3a] Soon after, they first demonstrated that the peptide-resin (PR) ester bond present in growing PRs made by Boc chemistry was susceptible to Ca^{2+}-promoted methanolysis, releasing the corresponding the *N*- and side-chain protected peptide methyl ester.[3b] In the following years, the group investigated Ca^{2+}- or Zn^{2+}-promoted peptide detachment from the Kaiser oxime resin (KOR) to give the corresponding *C*-TMPs. Examples included: i) transesterification of X-KOR bond from Ac-Ala-Gly-X-KOR (X: Gly, Ala, Phe or Lys (2-Cl-Z)) and from Ac-Ile-Ser(Bzl)-Asp(OZ)-KOR (Z: Bzl or cHex) to yield 85-96 % of the corresponding Ac-peptide-OMe;[4] ii) aminolysis of the Asp(X)-KOR bond from Ac-Ile-Ser(Bzl)-Asp(X)-KOR (X: OBzl or OcHex) to yield > 50 % of the corresponding Ac-peptide-amide;[4] iii) hydrolysis of X-KOR bond from Ac-Ile-Ser(Bzl)-X-KOR (X: Gly, Asp, Phe, Lys or Leu (2-Cl-Z)) to yield 19–90 % of the corresponding Ac-peptides-OH;[5] iv) thiolysis of the Asp(OcHx)-KOR bond from Ac-Ile-Ser(Bzl)-Asp(OcHx)-KOR to yield > 90 % of the corresponding Ac-peptide-SR.[6]

Objectives

With the intention of expanding this alternative single-step synthetic approach to SPPS-Fmoc chemistry, we have been studying peptide detachment from HMBA resins because the ester bond P-HMBA from these PRs is also susceptible to nucleophilic attack[7] and could be promoted by Ca^{2+}, Zn^{2+} or Cu^{2+} ions.

Experimental

The synthesis of aminoacyl (Fmoc-Asp(O^{t}Bu)-HMBA; Fmoc-Phe-HMBA, and Fmoc-Gly-HMBA) and PRs (Fmoc-Ile-Ser(But)-Asp(O^{t}Bu)-HMBA, Ac-Met-Ser-Leu-Phe-HMBA, and Fmoc-Gly-Gly-His(Trt)-Gly-HMBA) were performed using traditional or microwave-assisted synthesis at 60 °C and our customized protocols for Fmoc chemistry.[8] Their substitution levels (SL) were determined by total acidic hydrolysis of PRs at 130 °C for 24 h followed by amino acid analysis of the hydrolysates. The alcoholysis reactions were performed using Ca^{2+}, Zn^{2+} or Cu^{2+} ions as additive and RP-HPLC/LC-MS for monitoring. Identities of the *C*-esterified side-chain protected peptides formed were confirmed by electrospray ionization mass spectrometry (ESI-MS). Reaction yields were calculated from the difference between initial and final SLs.

Results and Discussion

Initially, we investigated alcoholysis of the ester bond of Fmoc-aminoacyl or Fmoc-peptidyl-resin varying the HMBA resins, the divalent metal ion (Ca^{2+} and Zn^{2+}), and the solvent system (20–50 % ROH/DMF; R: Et, ^{i}Pr and $(CH_2)_2CH_3$). The results showed that HMBA-AM and HMBA-PEGA resins were appropriate, $ZnCl_2$ was better metal salt than

$Ca(OAc)_2$ in 50% ROH/DMF at 60 °C for 48 h at molar ratio of 1:2 (peptide: metal ion). The Zn^{2+}-promoted alchoholyis of amino acid-resin ester bond from Fmoc-Gly-HMBA-AM produced the desired Fmoc-Gly-OR with yields > 97 % and methanolysis of the PR ester bond from Fmoc-Ile-Ser(^{t}Bu)-Asp(O^{t}Bu)-HMBA-AM furnished the desired product Fmoc-Ile-Ser(^{t}Bu)-Asp(O^{t}Bu)-OMe with yield of 87 %. All reactions were monitored by RP-HPLC; it was observed slow formation of the DBF (dibenzofulvene) resulting from removal of Fmoc group that led to little amounts of contamination. Aiming to overcome this undesirable side-reaction, the Fmoc group was replaced by acetyl group, so Zn^{2+}-promoted alcoholyses of the PR ester bond from Ac-Ile-Ser(^{t}Bu)-Asp(O^{t}Bu)-HMBA-AM using 50 % MeOH, EtOH, ^{i}PrOH, butan-1-ol or benzyl alcohol/DMF at 60 °C for 48 h formed the corresponding Ac-peptides-OR with yields of 73–95%, except for ^{i}PrOH that yielded only 20 % of the product due to sterical hindrance.

Afterwards, we studied the Zn^{2+}-promoted transesterification of Ac-Ile-Ser-Asp-OMe in Ac-Ile-Ser-Asp-OR (R:Et, But, Benzyl and ^{i}Pr)] under the same conditions. Such reactions produced the desired *C*-esterified peptides with yields of 73 – 92 %, except for that using ^{i}PrOH that formed the product with yield of 30 %. At last, to verify the tolerance of other functional groups, such as sulfur, in peptide sequences by our alternative method, we synthesized Ac-Met-Ser-Leu-Phe-HMBA, sequence analogous to that of the anti-inflammatory tripeptide For-Met-Ser-Leu-Phe.[9] Zn^{2+}-promoted methanolysis in 50 % MeOH/DMF, at 60 °C for 48 hfurnished the Ac-Met-Ser-Leu-Phe-OMe with yield of 93.5 % and high purity (83 %).

Currently, we are studying Shepherin I (Shep I)-derived PRs because this unique Gly- and His–rich peptide has 7 repetitions of the ATCUN motif GGH in its structure. which chelates Cu^{+2} and Ni^{2+} ions.[10] Once again, Cu^{2+}, Ca^{2+} and Zn^{2+}-promoted methanolyses of the PR ester bond from Fmoc-Gly-Gly-His(Trt)-Gly-HMBA-PEGA were effectively carried-out in 20% MeOH/DMF at 60 °C for 24 h to give the desired products with reaction yields >50 %.

In short, our alternative single-step method to produce *C*-TMPs based on divalent metal ion-promoted detachment of the peptide from resins showed to be compatible to Fmoc chemistry and HMBA resins, is efficient, quite selective, employs very mild conditions and neutral pH, being greener than the traditional.

Acknowledgments

The authors thank FAPESP (2022/01825-2) and CNPq (314470/2023-9) for the research grant and scholarship to MTM and TSF, Dr. Cleber W. Liria for the mass spectrometry and amino acid analyses and Matheus M. P. Landim for total acidic hydrolysis of PRs (undergraduate student).

References

1. Tanaka, S.; Kanno, M; Tashiro, Y. *Explor Drug Sci.* **2023**, *1*, 322.

2. a) Halpern, B.; Chew, L.; Close, V. et al. *Tetrahedron Lett.* **1968**, 5163. b) Weber, J. R.; Lokey, R. S. *Org. Letters.* **2010**, *12*, 1852.

3. a) Miranda, M. T. M.; Morita, H.; Tominaga, M. *Int. J. Peptide Prot. Res.* **1991**, *37*, 299. b) Miranda, M. T. M.; Theobaldo, F.C.; Tominaga, M. *Int. J. Peptide Prot. Res.* **1991**, *37*, 451.

4. Moraes, C.M., Bemquerer, M.P. & Miranda, M.T.M. *J. Peptide Res.* **2000**, *55*, 279.

5. Proti, P. B.; Remuzgo, C. Miranda, M. T. M. *J. Pept. Sci.* **2007**, *13*, 386.

6. Proti, P. B. & Miranda, M. T. M. *Tetrahedron Lett.* **2008**, *49*, 3853.

7. Alsina, J. & Albericio, F. *Biopolymers (Peptide Science)*, **2003**, *71*, 454-477.

8. Loffredo, C. et al. *J. Pept. Sci.* **2009**, *15*, 808.

9. Freire, T. S., Master Dissertation, Institute of Chemistry, University of São Paulo, **2013**.

10. a) Park. C. J. et al. *Plant Mol. Biol.* **2000**, *44*, 187. b) Remuzgo. C. et al. *Amino Acids.* **2014**, *46*, 2573. c) Nunes, L.G.P. et al., *Int. J. Pept. Res. Ther.* **2021**, *27*, 2945.

Proceedings of the 37th European Peptide Symposium
Michal Lebl, Editor
European Peptide Society, 2024 *http://doi.org/10.17952/37EPS.2024.P2104*

Glycine- and histidine-rich peptides derived from Shepherin I: study of candidacidal activity, metal-chelation properties and mode of action

Gustavo Nunes[1], Lohanna Lopes[1], Ernani Lacerda[2], Ana M. C. Ferreira[2], Carlos Marques[3], Breno P. Esposito[2] and M. Terêsa Machini[1]

Departments of Biochemistry[1] and Fundamental Chemistry[2], Institute of Chemistry, University of São Paulo, São Paulo, Brazil; [3]CNRS, Ecole Normale Supérieurie (ENS)[3], Lyon, France
gnunes@usp.br; mtmachini@iq.usp.br

Introduction

Antimicrobial Peptides (AMPs) have been considered very promising candidates for the development of new drugs capable of killing antibiotic-resistant pathogens, a global public health problem. Thus, it is important to explore the structure, properties and mode of actions of this class of bioactive peptides [1]. Glycine- and Histidine-Rich AMPs (GRH-AMPs) are rare in nature and very little studied. This small group includes Shepherin I (Shep I) [1,2], which is found in the roots of the Korean plant *Capsella bursa-pastoris* (*Brassicaceae* family) as a cationic product of *shep*-GRP polypeptide proteolysis [3] and contains 28 amino acid residues (67.9% glycine and 28.6 histidine) with 7 repetitions of the ATCUN motif GGH. In 2014, we reported that Shep I and fragments are more efficient in inhibiting the growth of *Candida* cells in the presence of $ZnCl_2$ [4].

Objectives

The present study is part of our ongoing research focused on Shep I. We developed and studied amidated Shep I and new short truncated analogues aiming to better understand Shep I´s structure-antifungal activity relationship, mode of action, and ability to chelate Zn^{2+} and Cu^{2+} ions [1,4]. The results could also inform on potential applications of these peptides in the fight against two major global public health problems: candidiasis and neurodegenerative diseases related to Cu^{2+}-promoted amyloid plaque formation [5].

Experimental and Results

Peptide synthesis employed microwave-assisted solid phase at 60 ºC and customized protocols. The purified peptides presented high purity degrees and the predicted amino acid contents. As expected, Shep Ia, truncated analogues, and their corresponding N-carboxyfluorescein (FAM)-forms inhibited the growth of *C. parapsilosis* ATCC 22019 (MICs of 25-100μM), *C. krusei* ATCC 6258 (MICs of 6.25-100 μM) and *C. albicans* ATCC 90028 (MICs ≥ 6.25 μM). Some of them also killed the fungal cells (MFCs of 6.25-100 μM) and became more potent in the presence of Zn^{2+} ions, confirming our previous observations [4]. Experiments using fluorescence spectroscopy, special metal-chelators and ESI-MS showed that the active FAM-peptides bind Cu^{+2} and Zn^{+2} ions with molar ratios of 1 Peptide:2 Cu^{2+} ions and 1 Peptide:10 Zn^{2+} ions. To speculate possible mode of actions for such efficient candidacidals, phase contrast and epifluorescence images were obtained using them and very simplified models of *Candida* plasma membrane (Giant Unilamellar Vesicles produced by *natural swelling* [6]); the results showed that the active fluorescent peptides strongly interact with the vesicles and penetrate and/or disrupt them after exposure of 20 min. *In vitro*, Shep I and new truncated analogues also competed with synthetic β-amyloid peptides for Cu^{+2} ions and prevented the formation of reactive oxygen species [7], findings highly suggestive of potential ability to interfere in Cu^{+2}-assisted amyloid plaques formation typical of neurodegenerative diseases. Preliminary *in vivo* experiments using FAM-octapeptide

http://doi.org/10.17952/37EPS.2024.P2104

showed penetration not only into *Candida* cells into *Caenohabditis elegans*, used as used as model organism for Blood Brain Barrier penetration [8].

Conclusions

C-amidated Shep I, some of the new short analogues and their corresponding FAM-forms are efficient candidacidals, potent chelators of divalent metal ions, and/or cell-penetrating peptides. Such functional versatility suggests that these Shep I-derived peptides find applications in science, human health and nutrition, agriculture, chemical and environmental activities.

Acknowledgments

The authors thank FAPESP (2022/01825-2) for the research grant to MTM, CNPq for research and doctoral scholarships to MTM (314470/2023-9) and LGPN (141836/2018-0), and to Dr. Cleber W. Liria for all mass spectrometry and amino acid analyses.

References

1. Nunes, L.G.P. et al. *Int J Pept Res Ther*. 2021, *27*, 2945.
2. Wang. G. et al. *Nucleic Acid Res*., 2016, *44*, 1087.
3. Park. C. J. et al. *Plant Mol. Biol*., 2000, *44*, 187.
4. Remuzgo, C. et al. *Amino Acids*., 2014, *46*, 2573; Reichert, T. Masters Dissertation, IQ-USP, 2018.
5. Oikonomou, G. & Shaham, S. *GLIA*, 2011, *59*, 9, 1253.
6. Conte-Daban et al. *Chemistry – A European Journal*, 2018, *24*, 33, 8447.
7. Dimova, R. & Marques, C. The Giant Vesicle Book (1st Ed.), CRC Press, p.676, 2019.
8. Kim, Y. et al. Physiol. Genomics, 2018, 50, 1002.

Proceedings of the 37th European Peptide Symposium
Michal Lebl, Editor
European Peptide Society, 2024 *http://doi.org/10.17952/37EPS.2024.P2111*

Molecular determinants of the anti-amyloid function of the cross-interaction of the human cathelicidin LL-37 with islet amyloid polypeptide

Valentina Armiento, Hendrik Wunderlich, and Aphrodite Kapurniotu
Division of Peptide Biochemistry, TUM School of Life Sciences, Technical University of Munich (TUM), Freising, D-85354, Germany

Introduction

Islet amyloid polypeptide (IAPP) is a 37-amino acid long neuroendocrine hormone which is secreted from the pancreatic β-cells [1]. IAPP is a highly amyloidogenic polypeptide. Its self-assembly into cytotoxic oligomers and amyloid fibrils is linked to inflammation, β-cell degeneration, and the pathogenesis of type 2 diabetes (T2D). Cathelicidins are a large group of multifunctional host-defense peptides (HDP) which are expressed by various cells including pancreatic β-cells and play important roles in inflammatory processes. The only known human cathelicidin is the 37-residue polypeptide LL-37. LL-37 has been long known for its broad-spectrum antimicrobial activity and its potent immunomodulatory effects [2]. Intriguingly, recent studies of our group showed that LL-37 is able to interact with IAPP with nanomolar affinity and to effectively suppress its amyloid self-assembly and related pancreatic β-cell-damaging effects *in vitro*. Our results suggested that the LL-37/IAPP interaction may play a protective role in T2D [3].

We asked which are the molecular and structural requirements of the anti-amyloid function of the LL-37/IAPP interaction. To address this, we first studied the effects of systematic N-terminal shortening of LL-37 on its structure, interactions with IAPP, and related anti-amyloid function.

Results and Discussion

Various different N-terminal truncated LL-37 analogues were synthesized by Fmoc-based solid phase peptide synthesis (SPPS) methodology. Their effects on fibril formation of IAPP were studied by using the amyloid specific thioflavin T (ThT) binding assay in combination with transmission electron microscopy (TEM). Their effects on formation of cytotoxic IAPP assemblies were determined using the MTT reduction assay in cultured rat insulinoma cells (RIN5fm). Of note, full length LL-37 was included in the studies for comparison.

We found that the deletion of the N-terminal dipeptide segment from LL-37 does not affect its ability to inhibit IAPP amyloid self-assembly and cytotoxicity. This suggests that residues Leu1 and Leu2 are not necessary for its anti-amyloid function. However, further N-terminal truncation up to LL-37(7-37) resulted in a loss of anti-amyloid activity. Thus, residues within LL-37(3-6) seem to play an important role in the anti-amyloid activity of LL-37. Interestingly, fluorescence spectroscopic titrations of fluorescently labeled IAPP with the truncated LL-37 analogues revealed that N-terminal truncation of LL-37 up to residue 14 does not impair binding affinity to IAPP.

Thus, the binding affinities of the LL-37 variants to IAPP did not correlate with their effects on IAPP amyloid formation suggesting that additional factors might be involved. Finally, far-UV CD spectroscopic studies were consistent with nearly identical secondary structure contents for LL-37 and LL-37(3-37) while further N-terminal truncation affected the secondary structure. The results of our studies should assist in deciphering the molecular basis of the anti-amyloid function of the LL-37/IAPP interaction and the design of novel molecules targeting pathogenic IAPP amyloid self-assembly in T2D.

Acknowledgments

This work was supported by Deutsche Forschungsgemeinschaft (DFG) grant SFB1035 B06 (part of SFB1035, project number 201302640).

References

1. Westermark, P., et al. *Physiol Rev* **91**, 795-826 (2011), https://doi.org/10.1152/physrev.00042.2009
2. Durr, U. H., et al. *Biochim Biophys Acta* **1758**, 1408-1425 (2006), https://doi.org/10.1016/j.bbamem.2006.03.030
3. Armiento, V., et al. *Angew Chem Int Ed Engl* **59**, 12837-12841 (2020), https://doi.org/10.1002/anie.202000148

https://doi.org/10.17952/37EPS.2024.P2116

Development of a potent and specific peptide inhibitor for the *E. coli* MazF toxin

Luiz R. Pizzolato-Cezar, Phelipe Vitale, Carolina D. Lacerda, Mario A. R. Pineda, Roberto K. Salinas, M. Terêsa Machini*

Department of Biochemistry, Institute of Chemistry, University of São Paulo, Av. Prof. Lineu Prestes, 748, 05508-900, São Paulo, Brazil

Corresponding author: mtmachini@iq.usp.br

Introduction

Antibiotics are compounds capable of interfering in active metabolic processes, such as transcription, translation, and cell wall synthesis. Therefore, in a population of pathogenic microorganisms, the cells that temporarily assume a dormant state are less susceptible to antibiotics. Cells displaying this phenotype are called persistent [1]. Although different pathways lead to this state, toxin-antitoxin systems (TAs) are assumed to play an important role in persistence [2].

TAs are abundant in the chromosome of prokaryotes, being composed of two types of protein: a toxin with cell growth inhibitory activity and a cognate antitoxin that keeps the toxin inactive under basal conditions. Through toxins, microorganisms enter a state of low metabolic activity and great resilience to different types of stress. Among the well-known TAs, a few have more prominent effects on bacterial physiology, such as the mazEF module. MazF is a 12 KDa-toxin with endoribonucleolytic action that inhibits cell growth through mRNA cleavage, while MazE is the cognate 9 KDa-antitoxin that keeps MazF inactive by blocking its catalytic site. MazEF complex, a heterohexamer, has a central role in several bacteria since it is involved in the formation of biofilm and increase of fitness in face of stress situations, such as phage infection, antibiotics, extreme temperatures and oxidative stress [3,4].

Inspired on such information, we designed this study aiming to discover a potent and specific peptide inhibitor for the toxin MazF [5].

Experimental

Briefly, peptide sequences of 12 amino acid residues interacting with *E. coli* His-MazF immobilised in a 96-well plate were selected with the Ph.D.TM-12 Phage Display Peptide Library kit (NEB, MA/USA). The best candidates were synthesized by microwave-assisted solid-phase synthesis at 60 °C using customized protocols [6], purified by RP-HPLC and characterized by ESI-MS and amino acid content. Cloning, expression, purification, and protein refolding technologies provided the recombinant *E. coli* MazE, His-MazF, and its mutant MazF(E24A). The synthetic peptides and MazE were compared in the capacity of inhibiting the endoribonuclease activity of His-MazF. Mapping of the interaction between the peptide inhibitor and the recombinant toxin or its mutant was performed by ITC, peptide analogues and NMR spectroscopy. 3D-structure analyses and calculations of toxin-P1 and toxin-antitoxin complexes by 800 MHz-NMR are in progress. Recently, effects of MazF inhibition on *E. coli* cells were also examined by bacterial growth, antibiotic surviving, oxygen consumption and resazurin assays [5,7].

Results and Discussion

The results obtained so far demonstrate that a potent and specific peptide inhibitor for *E. coli* MazF toxin (Fig. 1A) was discovered and studied. P1 contains 12-amino acid residues and acts as a reversible competitive inhibitor (Fig. 1B). Extensive scanning of P1 amino acid sequence using different techniques and a few synthetic analogues allowed mapping the interactions of the synthetic P1 with the recombinant His-MazF and comparing the points of contact with those already described for the MazEF complex. P1 proved to be also functional *in vivo* as it prevented the deterministic activation of the toxin that causes a strong metabolic inhibition with consequent acquisition of bacterial tolerance to antibiotics. Overall, these results indicated that MazF is a possible regulator of bacterial persistence.

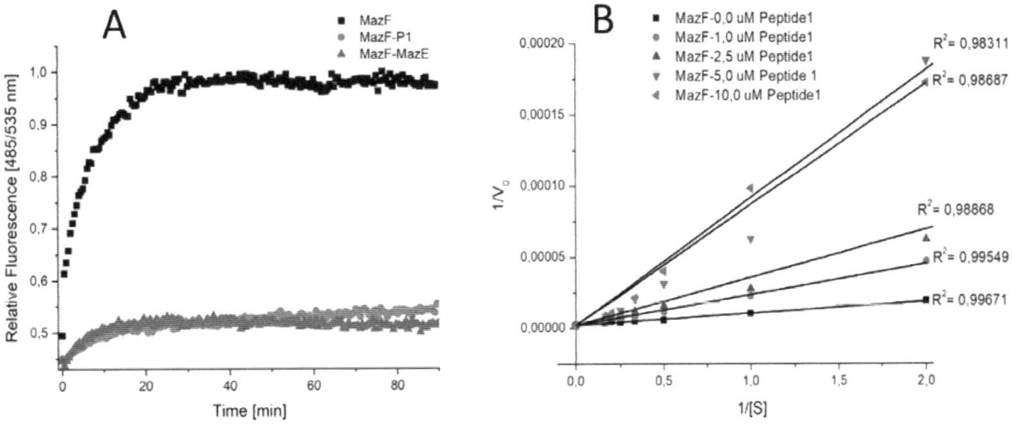

Fig. 1. *In vitro* catalytic profiles of *E. coli* His-MazF using a specific fluorescent substrate in the absence and presence of MazE or the synthetic peptide (**A**). Plot of Lineweaver-Burk in the presence of P1(**B**).

Acknowledgments

The authors thank FAPESP (2022/01825-2) for the research grant to MTM, CNPq for research and doctoral scholarships to MTM (314470/2023-9) and LRPC (141141/2016-6), respectively, Dr. Cleber W. Liria for all mass spectrometry and amino acid analyses, Prof. German Sgro (FCFRP-USP) for discussion and help with protein cloning, expression and purification, Jose D. Rivera and A. Davalos for help with 800-MHz NMR analyses, Profs. S.R. Marana, I.M. Cuccovia, S.C. Farah, C.O. Columbano, and A. Kowaltosvski (IQ-USP) for advices and/or sporadic use of their equipments.

References

1. Kumar, A. Rahal, J. S. Sohal, V. K. Gupta, *Expert Rev. Anti. Infect. Ther.*, 19, 2, 121–127, 2021.
2.Singh, G.; Yadav, M.; Ghosh C.; Rathore, J. S. *Curr. Res. Microb. Sci.*,2, 1000472, 2021
3. Pizzolato-Cezar, LR; Spira, B.; Machini, M.T. *Curr. Res. Microb. Sci.*, 5 100204, 2023
4. Al-Hinai et al. Appl. Environ. Microbiol. 78(22), 8112-8121, 2012
5. Loffredo, C.; Assunção, N.A.; Miranda, M.T.M. *J. Pept. Sci.*, *15*, 808-817, 2009.
6. Pizzolato-Cezar, L.R. Ph.D. Thesis, Institute of Chemistry, University of São Paulo, 2023
7. Pizzolato-Cezar, L.R. *et al.* Development of a potent and functional *in vivo* inhibitor peptide for the toxin MazF (*in preparation*)

Proceedings of the 37th European Peptide Symposium
Michal Lebl, Editor
European Peptide Society, 2024
https://doi.org/10.17952/37EPS.2024.P2169

Design and characterization of a multistage peptide-based vaccine platform to target *Mycobacterium tuberculosis* infection

C. Bellini[1,2,‡], E. Vergara3, F. Bencs[4], Sz. Bősze[5], D. Krivić[6], B. Bacsa[6], S.E. Surguta[7], J. Tóvári[7], R. Reljic[3], and K. Horváti[1*]

[1]*MTA-HUN-REN TTK Lendület Peptide-Based Vaccines Research Group, Institute of Materials and Environmental Chemistry, HUN-REN Research Centre for Natural Sciences, Budapest, Hungary, *email address: horvati.kata@ttk.hu*
[2]*Hevesy György PhD School of Chemistry, Eötvös Loránd University, Budapest, Hungary;*
[3]*Institute for Infection and Immunity, St. George's, University of London, London, UK;*
[4]*Laboratory of Structural Chemistry and Biology, Institute of Chemistry, Eötvös Loránd University, Budapest 1117, Hungary*
[5]*HUN-REN-ELTE Research Group of Peptide Chemistry, Budapest, Hungary;*
[6]*Division of Medical Physics and Biophysics, Gottfried Schatz Research Center, Medical University of Graz, Graz, Austria;*
[7]*Department of Experimental Pharmacology and National Tumor Biology Lab., National Institute of Oncology, Budapest, Hungary.*
[‡]*Present address: Interdepartmental Laboratory of Peptide and Protein Chemistry and Biology, Department of NeuroFarBa, University of Florence, Sesto Fiorentino (FI), 50019, Italy.*

Tuberculosis (TB) is the second leading cause of death by a single infectious agent worldwide. The Bacille Calmette-Guérin (BCG), the only available vaccine against TB, provides inconsistent protection against pulmonary TB in adults, the primary source of disease transmission. Thus, a more effective vaccine is urgently needed to stop the TB epidemic. The development of a new vaccine is hindered by the complex immunopathology of *Mycobacterium tuberculosis* (*Mtb*), especially in eliciting protection against both active and latent stages of infection.

Multistage subunit vaccines, which incorporate antigens expressed in both phases, are designed to address different stages of a pathogen's life cycle or diverse aspects of the immune response in one construct. This strategy may be particularly advantageous when dealing with complex pathogens characterized by high antigen variability or propensity for mutagenesis, such as *Mtb*. However, their main limitation is that they often elicit a less robust immune response compared to traditional vaccines. In addition, they may face challenges related to bioavailability, including rapid clearance or undesired post-injection degradation. Consequently, developing subunit vaccines requires a well-designed formulation, incorporating immunostimulatory adjuvants and a delivery system to address these limitations [**1**].

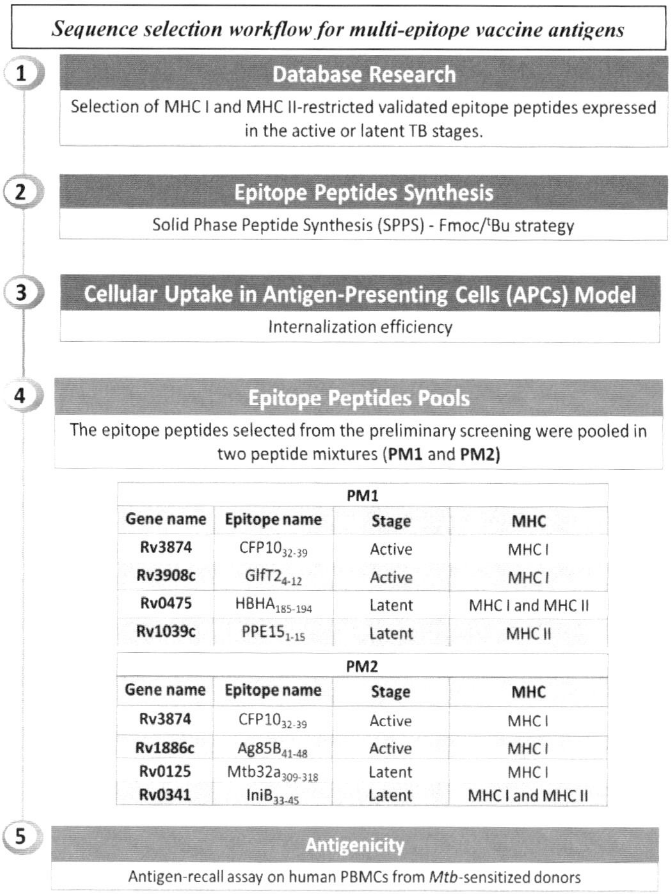

PM1			
Gene name	**Epitope name**	**Stage**	**MHC**
Rv3874	$CFP10_{32\text{-}39}$	Active	MHC I
Rv3908c	$GlfT2_{4\text{-}12}$	Active	MHC I
Rv0475	$HBHA_{185\text{-}194}$	Latent	MHC I and MHC II
Rv1039c	$PPE15_{1\text{-}15}$	Latent	MHC II

PM2			
Gene name	**Epitope name**	**Stage**	**MHC**
Rv3874	$CFP10_{32\text{-}39}$	Active	MHC I
Rv1886c	$Ag85B_{41\text{-}48}$	Active	MHC I
Rv0125	$Mtb32a_{309\text{-}318}$	Latent	MHC I
Rv0341	$IniB_{33\text{-}45}$	Latent	MHC I and MHC II

Epitope selection:
A set of validated T cell epitopes was chosen from the available literature and synthesized using solid-phase peptide synthesis. After the screening, the 4 epitopes contained in the peptide mixture PM1 were selected for the inclusion in the vaccine platform.

Lipopeptide-based vaccine platform: The peptides included in PM1 were used to synthesize the peptide-based vaccine platforms. $CFP10_{32-39}$, $GlfT2_{4-12}$, and $HBHA_{185-194}$ were combined in one linear peptide chimera, which was derivatized with a maleimide moiety by orthogonal reaction. The $PPE15_{1-15}$ was elongated with a cysteine residue at the *C*-terminal and then conjugated to the linear peptide through the thiol-maleimide reaction. The *N*-term of the peptide chimera was either acetylated or palmitoylated to obtain the final branched conjugates (Ac-CGHP and Pal-CGHP). Palmitoylation previously demonstrated to improve the cellular uptake and the immunogenicity of peptide-based vaccines [**2**].

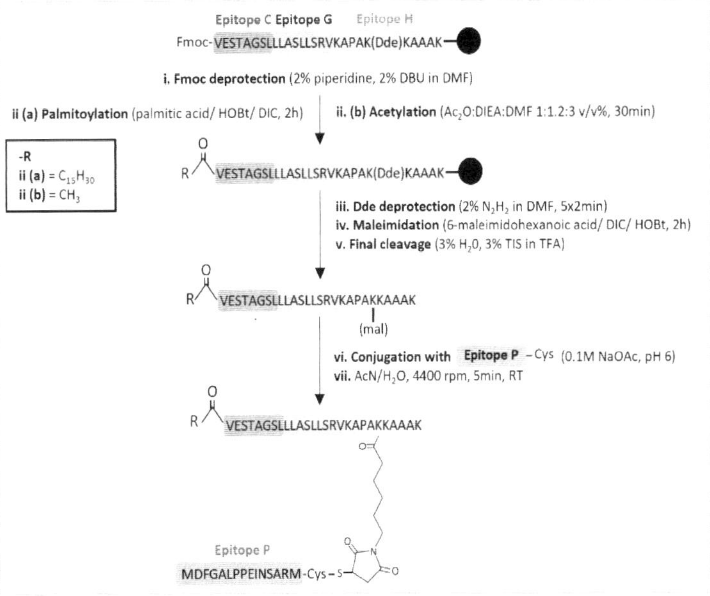

Immunological evaluation.

The Pal-CGHP conjugate, the peptide mixture PM1 in PBS, and PM1 formulated with the Sigma Adjuvant System (SAS) were administered subcutaneously to BALB/c mice three times, two weeks apart. We then measured antigen-specific splenic T cell proliferation and cytokine secretion among immunised and unimmunised animals. As expected, PM1 did not produce a noticeable immune response, likely due to intrinsic poor immunogenicity and *in vivo* stability of short synthetic peptides. Pal-CGHP immunization produced significant T cell proliferation, which slightly increased over time, with the highest levels observed four weeks post-immunization. Similarly, IFNγ and TNFα concentrations followed a time-dependent increase. In contrast to the SAS-formulated candidate, Pal-CGHP induced significant IL-22 and IL-10 production but no IL-17 response. When splenocytes from Pal-CGHP immunized mice were restimulated with the individual C/G/H/P peptides, bearing the original sequences, epitope-specific antigenicity was observed, suggesting that their chemical modification occurred in the conjugation did not compromise the epitope recognition. Thus, the results suggest that Pal-CGHP is more promising than the two tested candidates and it merits further investigation as a multi-epitope self-adjuvating vaccine platform [**3**].

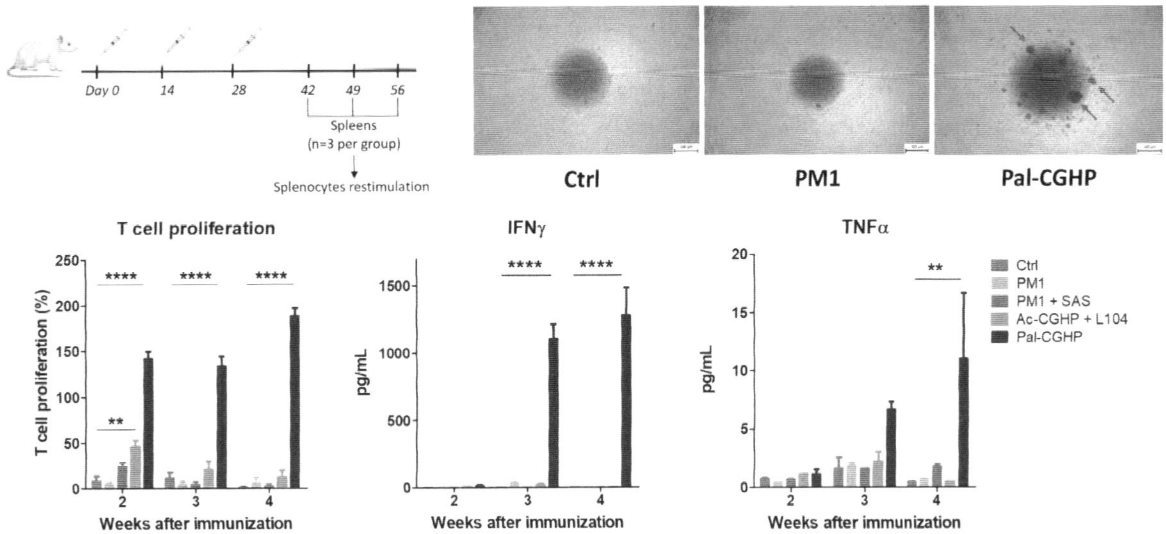

Acknowledgements

Authors thank the European Union's H2020-MSCA-ITN programme BactiVax (No. 860325) and the Momentum Program (LP2021-28) of the Hungarian Academy of Sciences.

References

[**1**] *Bellini C, Horváti K. Cells. 2020, 2673;* [**2**] *Horváti, K. et al. Vaccines. 2019, 7, 101;* [**3**] *Bellini C. et al. Bioconjugate Chem. 2023, 34, 10, 1738–1753*

Proceedings of the 37th European Peptide Symposium
Michal Lebl, Editor
European Peptide Society, 2024
http://doi.org/10.17952/37EPS.2024.P2224

Synthesis and characterization of nucleolin receptor recognizing F3 peptide fragment-drug conjugates

Bettina Basa[1, 2, 3], Diána Vári-Mező[3, 4, 5], Kata Nóra Enyedi[1, 3], József Tóvári[5], Szilvia Bősze[3, 6], Krisztina Kiss[3], Gábor Mező[1, 3]

[1]*ELTE Eötvös Loránd University, Faculty of Science, Institute of Chemistry, Department of Organic Chemistry, Budapest, 1117 Hungary,* [2]*ELTE Hevesy György PhD School of Chemistry, Budapest, 1117 Hungary,* [3]*HUN-REN-ELTE Research Group of Peptide Chemistry, Budapest, 1117 Hungary,* [4]*School of Ph.D. Studies, Doctoral School of Pathological Sciences, Semmelweis University, Budapest, 1085 Hungary;* [5]*Department of Experimental Pharmacology and the National Tumor Biology Laboratory, National Institute of Oncology, Budapest, 1122 Hungary;* [6]*Department of Genetics, Cell- and Immunobiology, Faculty of Medicine, Semmelweis University, Budapest, 1089 Hungary*

Introduction

Chemotherapy remains the most widely used approach in cancer treatment. Despite its widespread use, the frequency of tumor relapse, and the associated severe toxic side effects, require continuous methodological development. Targeted tumor therapy has emerged as a promising route for this development. This strategy involves conjugating drugs with targeting molecules that exhibit high specificity and selectivity towards tumor-selective proteins or receptors. By doing so, the drug delivery becomes tailored to tumor cells, minimizing the impact on healthy cells [1]. Nevertheless, a pervasive challenge in targeted therapeutic methods is the limited number and types of tumor-specific markers. This is particularly true for tumor subtypes, called tumor stem-like cells (CSCs) [2], where the lack of overexpression of common tumor markers impedes methodological progress. In addition, because of their high self-renewal properties, CSCs are also implicated in cancer relapse. Thus, the development of effective compounds for targeted tumor therapy is necessary that can recognize not only the conventional tumor cells but the resistant CSCs, too. Nucleolin is a protein that is expressed both on sensitive cells and resistant CSCs, which makes it a suitable target in tumor therapy. The 31-mer peptide (F3), selectively binds to the nucleolin receptor, offering a novel approach for targeted tumor therapy [3]. This research endeavors to optimize the F3 peptide structure for the development of a more cost-effective drug candidate.

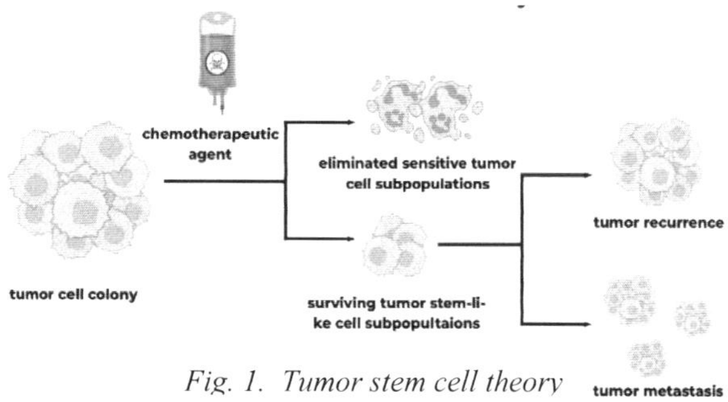

Fig. 1. Tumor stem cell theory

Results and Discussion

In this study, we aimed to develop peptide-daunomycin conjugates with enhanced targeting activity. To ensure uniform drug release we inserted the cathepsin-B enzyme labile GFLG spacer between the peptide and daunomycin. Daunomycin was selected as the chemotherapeutic agent due to its intrinsic autofluorescence, which allowed for simultaneous evaluation of cellular uptake and cytostatic effect. The efficacy of the conjugates was measured using MTT assays on human melanoma (A2058) and human colorectal adenocarcinoma (HT-29) cell lines. To optimize the peptide sequence for nucleolin receptor binding, we synthesized eight overlapping decapeptides that cover the entire F3 peptide. The *in vitro* cytostatic activity studies demonstrated that the peptide-daunomycin conjugate, that contains the first 10 amino acids of the N-terminal of the F3 peptide, exhibited higher antitumor activity compared to the full-length conjugate. To identify the shortest, most effective targeting sequence we synthesized peptides containing the first 7,13, and 20 amino acids of the N-terminal. The cytotoxicity results indicated, that shortening the sequence reduced the efficacy of the conjugate, while elongating the sequence did not improve the effectiveness.

Table 1. – Length optimization of the targeting sequence

Sequence	Cytostatic effect IC_{50} (μM)	
	A2058	*HT-29*
Dau-Aoa-GFLG-^{1}KDEPQRRSARLSAKPAPPKPEPKPKKAPAKK^{31}-NH_2	2,99 ± 0,37	11,78 ± 0,12
Dau-Aoa-GFLG-^{1}KDEPQRRSARLSAKPAPPKP^{20}-NH_2	3,35 ± 0,12	87,85 ± 0,13
Dau-Aoa-GFLG-^{1}KDEPQRRSARLSA^{13}-NH_2	3,60 ± 0,09	90,91 ± 0,15
Dau-Aoa-GFLG-^{1}KDEPQRRSAR^{10}-NH_2	3,78 ± 0,12	28,32 ± 0,09
Dau-Aoa-GFLG-^{1}KDEPQRR^{7}-NH_2	17,15 ± 0,10	> 100

Further optimization was carried out using alanine scanning, in which each amino acid was systematically substituted with apolar alanine, to evaluate its role in the conjugate's activity. This approach identified critical residues, like the DEPQ sequence and the C-terminal arginine, whose modification significantly diminished the conjugate's efficacy, which means these amino acids could be important in the attachment to the nucleoline receptor. Conversely, substitutions at the N-terminal lysine, and the 8th position serine improved the conjugate effectiveness, suggesting that these positions tolerate modifications.

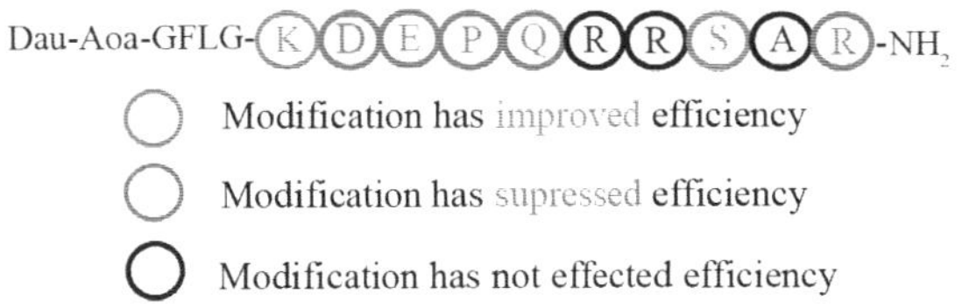

Fig. 2. Results of the alanine scan

Table 2. – Alanine scan

Sequence	*Cytostatic effect* *IC$_{50}$ (µM)*	
	A2058	*HT-29*
Dau-Aoa-GFLG-^{1}KDEPQRRSAA^{10}-NH$_2$	32,0 ± 3,4	> 50
Dau-Aoa-GFLG-^{1}KDEPQRRAAR^{10}-NH$_2$	11,0 ± 1,2	17,2 ± 9,1
Dau-Aoa-GFLG-^{1}KDEPQRASAR^{10}-NH$_2$	22,8 ± 2,8	26,7 ± 3,6
Dau-Aoa-GFLG-^{1}KDEPQARSAR^{10}-NH$_2$	15,1 ± 1,0	33,0 ± 16,9
Dau-Aoa-GFLG-^{1}KDEPARRSAR^{10}-NH$_2$	> 50	> 50
Dau-Aoa-GFLG-^{1}KDEAQRRSAR^{10}-NH$_2$	> 50	> 50
Dau-Aoa-GFLG-^{1}KDAPQRRSAR^{10}-NH$_2$	39,1 ± 8,4	> 50
Dau-Aoa-GFLG-^{1}KAEPQRRSAR^{10}-NH$_2$	> 50	> 50
Dau-Aoa-GFLG-^{1}ADEPQRRSAR^{10}-NH$_2$	3,3 ± 0,3	9,7 ± 3,4

Additionally, to improve both the efficacy and serum stability of the conjugates, non-natural amino acids and intramolecular cyclization were introduced. Although the incorporation of a disulfide bridge slightly reduced the cytotoxicity of the peptide-drug conjugate, it could increase the serum stability, ultimately enhancing the overall therapeutic potential of the conjugate. The preliminary in vivo experiments confirmed that the shorter peptides with optimized sequences might be better drug conjugates than the F3-drug conjugate.

Table 3. – Incorporation of non-natural amino acids and intracellular rings

Sequence	*Cytostatic effect* *IC$_{50}$ (µM)*	
	A2058	*HT-29*
Dau-Aoa-GFLG-^{1}KDEPQRCitSAR^{10}-NH$_2$	19,70 ± 0,10	> 50
Dau-Aoa-GFLG-^{1}KDEPQCitRSAR^{10}-NH$_2$	18,20 ± 0,14	> 50
Dau-Aoa-GFLG-^{1}NleDEPQRRSAR^{10}-NH$_2$	20,30 ± 0,07	> 50
Dau-Aoa-GFLG-^{1}KDEPQF$_{(pCl)}$RSAR^{10}-NH$_2$	10,13 ± 0,15	> 50
Dau-Aoa-GFLG-^{1}CDEPARRCAR^{10}-NH$_2$ (-S---------S-)	8,17 ± 0,11	30,23 ± 0,07
Dau-Aoa-GFLG-^{1}KDEAQRRCAR^{10}-NH$_2$ (HN-CO-CH$_2$-S)	12,94 ± 0,12	> 50

Acknowledgments

The research was supported by the National Research, Development and Innovation Office under grants NKFIH K146039.

References

1. G. Mező, M. Manea - Receptor-mediated tumor targeting based on peptide hormones - Expert Opinion in Drug Delivery, 2010; 7; 79-96; https://doi.org/10.1517/17425240903418410
2. L. S. Carvalho, N. Grançalves, N. A. Fonseca, J. N. Moreira: Cancer Stem Cells and Nucleolin as Drivers of Carcinogenesis - Pharmaceuticals - 2021; 14(1); 60; 1-21; https://doi.org/10.3390/ph14010060
3. K. Prokka, P. Laakkonen, J. A. Hoffman, M. Bernsconi, E.Ruoslathi. - A fragment of the HMGN2 protein homes to the nuclei of tumor cells and tumor endothelial cells in vivo - Proceedings of the National Academy of Sciences of the United States of America, Biochemistry, 2002, 99, 7444-7449 https://doi.org/10.1073/pnas.062189599

Proceedings of the 37th European Peptide Symposium
Michal Lebl, Editor
European Peptide Society, 2024 *10.17952/37EPS.2024.P2236*

Non-ionic surfactants for the solubilisation of lipopeptides and enhancement of their biological activity

Ágnes Ábrahám [a,b], Gergő Gyulai [a,b], Judith Mihály [c], Andrea Horváth [d], Orsolya Dobay [d], Éva Kiss [b] and Kata Horváti [a]*

[a] *MTA–HUN-REN TTK Lendület "Momentum" Peptide-Based Vaccines Research Group, Institute of Materials and Environmental Chemistry, Research Centre for Natural Sciences, Budapest, Hungary,* *email address: horvati.kata@ttk.hu
[b] *Laboratory of Interfaces and Nanostructures, Eötvös Loránd University, Budapest, Hungary*
[c] *HUN-REN TTK Biological Nanochemistry Research Group, Institute of Materials and Environmental Chemistry, Research Centre for Natural Sciences, Budapest, Hungary*
[d] *Institute of Medical Microbiology, Semmelweis University, Budapest, Hungary*

Lipopeptides have been extensively explored for their antimicrobial, anticancer and hormone activity, as well as promising fully synthetic self-adjuvanted vaccine candidates. However, lipopeptides tend to violate classical medicinal chemistry filters, mainly due to their high molecular weight and LogP characteristics. Lipopeptides usually form heterogenic aggregates in the micrometre range, therefore filtration is not an appropriate method to ensure sterility of the suspension. In addition, low aqueous solubility is a challenge when performing cell-based assays and often leads to failure in drug development. Besides, the often occurring haemolytic and cytotoxic side effects of lipopeptides may also be a serious limitation for their *in vitro* and *in vivo* use.

Table 1. *Modell lipopeptides selected for this study*

Name	Composition	Molecular weight	Net charge	Biological activity
pL1	palmitoyl-LLRK	766.1	2	Short antibacterial lipopeptide
pCM15	palmitoyl-KWKLFKKIGAVLKVL	2008.6	5	Hybrid peptide with anticancer activity
pATIPC	palmitoyl-KLVANNTRLTKPK-(FFGINTIPIAC)-GGLIDIAPHQISSV	4496.8	4	Branched-chained vaccine antigen

Non-ionic surfactants: poloxamers

In this study, self-assembling triblock copolymers, namely poloxamers were investigated to control the size of aggregates and i*n vitro* properties of biologically active lipopeptides (***Table 2***). Poloxamers are non-ionic amphiphilic copolymers consisting of poly(ethylene-oxide)—poly(propylene-oxide)—poly(ethylene-oxide) blocks. Their versatility arises from their tunable properties, which are linked to the length of their hydrophilic and hydrophobic chains. They are applied in various fields, serving as drug matrices and carriers, surfactants, solubilizers, wetting agents, emulsifiers, and adsorption enhancers. Due to their biocompatibility the U.S. Food and Drug Administration (US FDA) and European Medicines Agency (EMA) have approved numerous poloxamers for human use. Moreover, polymer micelles, which are more resistant to dilution than conventional surfactants, can be produced on a large scale. Poloxamers can encapsulate molecules within the core, constructed by the hydrophobic PPO segment, while the hydrophilic PEO region acts as a protective shield. The formed micelles have small size (10-100 nm), and exhibit spherical or rod-like shapes. In this study, the below listed poloxamers were investigated and the Critical Micelle Concentration (CMC) together with the cytotoxicity effect was determined. Based on our results, three poloxamers were selected (P-104, P-123, F-127) that have relatively low CMC values and can be used in high concentrations without toxicity.

Table 2. Characteristics of the used poloxamers

Poloxamer name	Pluronic name	Molecular weight	EO units	PO units	HLB values	CMC literature (g/L)	CMC experimental (g/L) [1]	Cytotoxicity IC50 (g/L) [2]
334	P-104	5900	27	56	13	0.02-0.08	0.02	2.731
403	P-123	5800	20	70	8	0.01-0.02	0.02	>2.9
407	F-127	12600	97	63	22	0.03-0.2	0.08	>6.3

The state of the poloxamer is marked with liquid (L), flakes (F), and paste (P). EO: ethylene oxide; PO: propylene oxide.HLB: hydrophilic-lipophilic balance.
[1] Critical micelle concentration (CMC) values of the poloxamers at 37°C in PBS, determined by pyrene fluorescence probe method
[2] Half-maximal inhibitory concentration (IC_{50}) values of the poloxamers, measured on HepG2 human cell-line, using AlamarBlue viability assay.

Characteristics of the poloxamer – lipopeptide mixed micelles

Mixed micelles of poloxamers with model lipopeptides were prepared by a simple solvation method and through dynamic light scattering (DLS) measurements, the production of small particles (diameter of around 20 nm) and reproducible, storable systems were demonstrated. Besides, these micelles can also tolerate freeze-thaw and lyophilisation-resolution cycles.

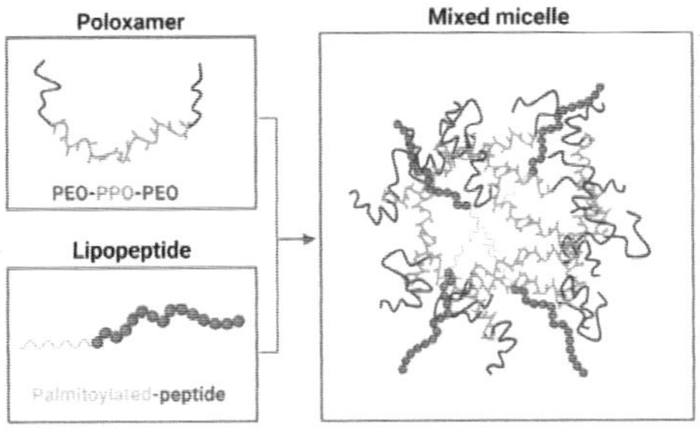

Characteristics of the mixed micelles:

Lipopeptide	Encapsulation efficacy	d_H (nm)	PDI
pL1	93.3 %	18.7 ± 0.3	0.07 ± 0.05
pCM15	76.5 %	22.7 ± 0.2	0.25 ± 0.03
pATIPC	60.4 %	19.8 ± 0.2	0.15 ± 0.02

Stability of the mixed micelles:

	pL1	pCM15	pATIPC
	d_H (nm)	d_H (nm)	d_H (nm)
Stored at RT for one week	21.9 ± 2.1	26.9 ± 0.7	30.3 ± 0.7
Lyophilized-resuspended	19.2 ± 0.3	19.5 ± 0.2	19.6 ± 0.2
Frozen -thawed	18.2 ± 0.7	19.6 ± 0.2	19.2 ± 0.3

Improved effectivity and selectivity

During the biological assessments, the properties of lipopeptide and lipopeptide–poloxamer systems were compared. The results showed that selectivity of pL1 lipopeptide was significantly improved by the poloxamers, increasing its antibacterial activity while reducing its haemolytic activity. The cytostatic effect of pCM15 colorectal tumor cells in the presence of poloxamers was approximately twice as high as that of the unformulated lipopeptide. For the self-adjuvanted vaccine antigen pATIPC, the rate of cellular uptake by antigen presenting model cells increased several-fold in the presence of poloxamers. Among the poloxamers, the highest biological activity was observed when Plur123 was used.

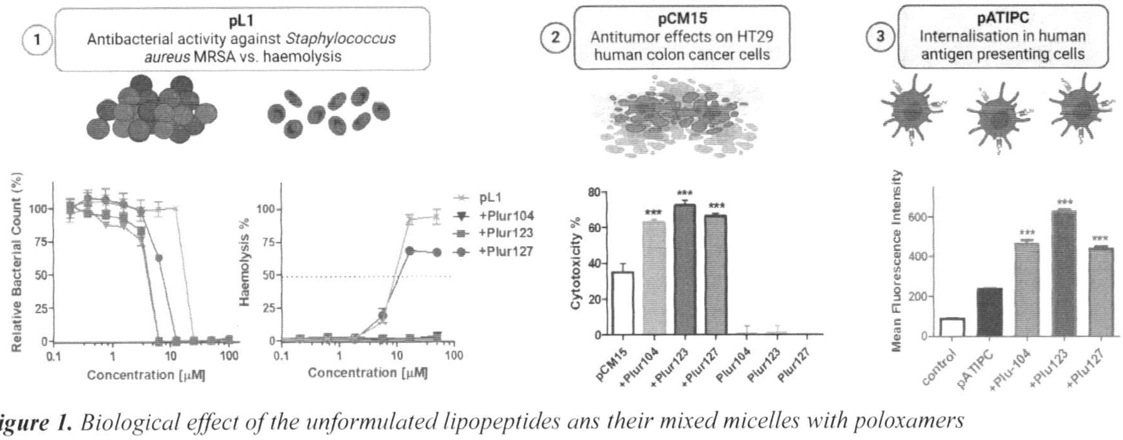

Figure 1. *Biological effect of the unformulated lipopeptides ans their mixed micelles with poloxamers*

Overall, it is concluded that poloxamer formulation is effective in optimizing the effect of lipopeptides, but the final properties of the mixed micelles are highly dependent on the length and structure of the peptide chain. It is believed that the optimal peptide chain length that can still be hidden in the PEO chains of Plur123 is around 10–12 amino acids.

Acknowledgements

This work was supported by the Lendület (Momentum) Programme of the Hungarian Academy of Sciences (LP2021-28).

Reference

Ábrahám, Á. et al. Optimizing lipopeptide bioactivity: The impact of non-ionic surfactant dressing. *J Pharm Analysis*, 2024, 101020.

 http://doi.org/10.17952/37EPS.2024.P2241

Novel viral peptide-based tandemized carriers: Structure prediction and *in vitro* evaluation

Katalin Uray[1], Martin Krátký[2], Jiří Horčic[2], Levente Lakatos[1,3], Előd Méhes[4], Bálint Jezsó[5,6], Jarmila Vinšová[2], Szilvia Bősze[1,7]

1 HUN-REN–ELTE Research Group of Peptide Chemistry, Hungarian Research Network, Eötvös Loránd University, H-1117 Budapest, Hungary, 2 Department of Organic and Bioorganic Chemistry, Faculty of Pharmacy, Charles University, CZ-50005 Hradec Králové, Czech Republic, 3 Doctoral School of Biology, Institute of Biology, Eötvös Loránd University, H-1117 Budapest, Hungary,4 Institute of Physics, Department of Biological Physics, Eötvös Loránd University, H-1117 Budapest, Hungary, 5 "Momentum" Motor Enzymology Research Group, Department of Biochemistry, Eötvös Loránd University, H-1117 Budapest, Hungary, 6 HUN-REN Research Centre for Natural Sciences, H-1117 Budapest, Hungary, 7 Department of Genetics, Cell- and Immunobiology, Semmelweis University, Faculty of Medicine, H-1089 Budapest, Hungary

Introduction

Previously we have shown the efficacy of a modified synthetic peptide (RF) derived from the Herpes simplex virus-1 gD glycoprotein (^{228}QRTVAVYSLRIAGFHG^{243}) (Fig. 1) ([1] in internalization into epithelial or neural cells. Tuftsin derivatives (OT5(P), OT5) can deliver cargo into different cells via neuropilin receptors [2-4]. Positively charged penetratin, the most studied cell-penetrating peptide (Pen), can rapidly enter cells.
We aimed to modify the RF peptide's cellular uptake rate and intracellular localization by tandemizing it with tuftsin or penetratin derivatives (Table 1, Fig. 2).
Internalization was evaluated on VERO E6 non-human primate kidney epithelial and A431 human skin squamous carcinoma cells.

Results and Discussions

Internalization of synthetic Cf-labeled peptides and tandems was measured by BD LSR II flow cytometer. All Cf-peptides containing the RF sequence internalize in a concentration-dependent manner (UC_{50} concentration value for Cf-positivity in 50% of the live cells: 0.1 – 5 μM) (Figure 2, Table 1). No cytotoxicity was detected (Figure 2, inserts). According to confocal laser scanning microscopy Cf-RF showed cytoplasmic localization; its tandemization with the tuftsin derivative resulted in compartmentalized localization to different degrees (data not shown).

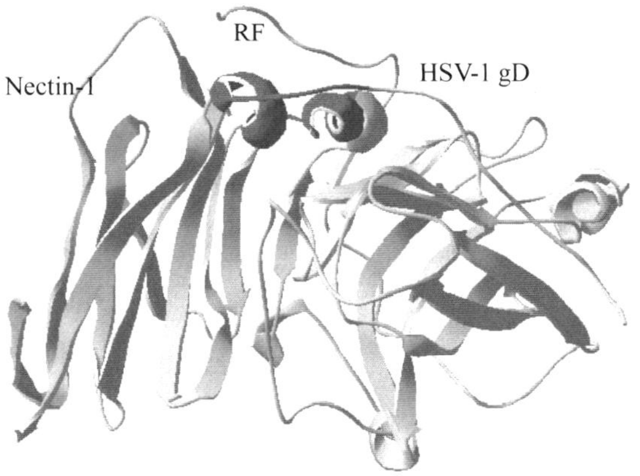

Figure 1. Secondary structure of HSV-1 gD–nectin-1 cell adhesion complex (PDB: 3U82) for virus entry. HSV gD 228-243 region show cellular entry ability as a synthetic peptide (Cf-RF).

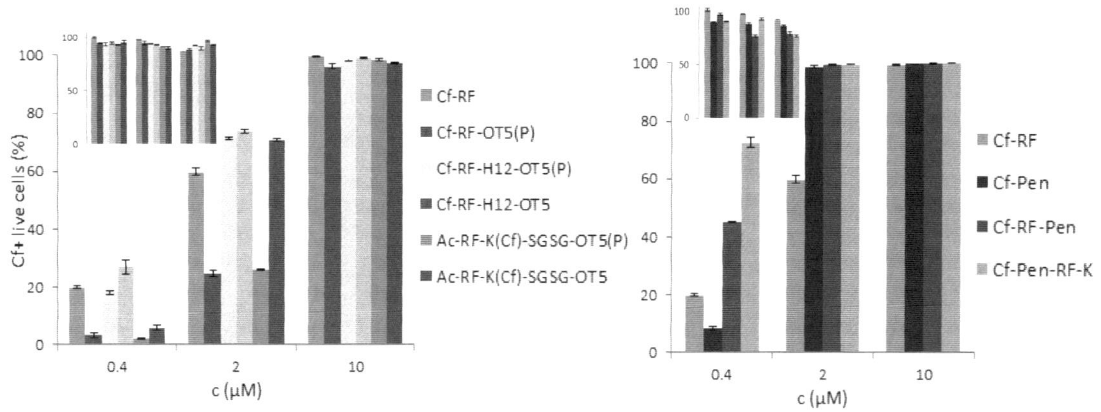

Figure 2. Internalization of Cf-peptides into A431 epithelial cells (B), represented by Cf+ live cell percentage. Inserts: relative viability (%).

Table 1. Analytical data, and internalization of Cf-labeled peptides characterized by UC_{50} (concentration value for Cf-positivity in 50% of the live cells)

Peptide code	*Mw, average*[a]		*UC_{50} (μM)*[b]	
	calc.	*meas.*	*A431*	*VERO E6*
Cf-RF	2132.36	2132.17	1.34	4.72
Cf-RF-OT5(P)	2712.10	2710.81	3.54	1.36
Cf-RF-H12-OT5(P)	3859.31	3857.33	1.04	1.08
Cf-RF-H12-OT5	3791.23	3790.98	0.88	2.37
Ac-RF-K(Cf)-SGSG-OT5(P)	3170.49	3169.44	3.41	0.51
Ac-RF-K(Cf)-SGSG-OT5	3102.41	3101.13	1.19	1.23
Cf-Pen	2605.06	2603.08	0.84	0.10
Cf-RF-Pen	4361.11	4361.42	0.46	0.60
Cf-Pen-RF-K	4489.28	4489.82	0.24	0.54

[a] *ESI-MS (Bruker Amazon SL ESI Mass Spectrometer, Bremen, Germany)*

[b] *Concentration value for Cf-positivity in 50% of the live cells (BD LSR II flow cytometer, data evaluation: FACSDiva 5.0 software.*

The secondary structure of tandem peptides (Pep-Fold 3) [5] depends on the linker moiety. The H12 rigid spacer may help the RF region to adopt its native helical structure, thus helping internalization in VERO E6 cells. Penetratin derivatives and tandems internalize the most effectively; in A431 cells, nectin-binding HSV gD RF peptide enhances the internalization. Different spacers and tuftsin vs. tuftsin agonists in the tandemized peptides showed opposite orders of internalization in A431 and VERO E6 cells. Internalization into VERO E6 cells is favored by the presence of linker, preferably a shorter, more flexible one. On the other hand, in A431 cells the rigid linker proved more effective.

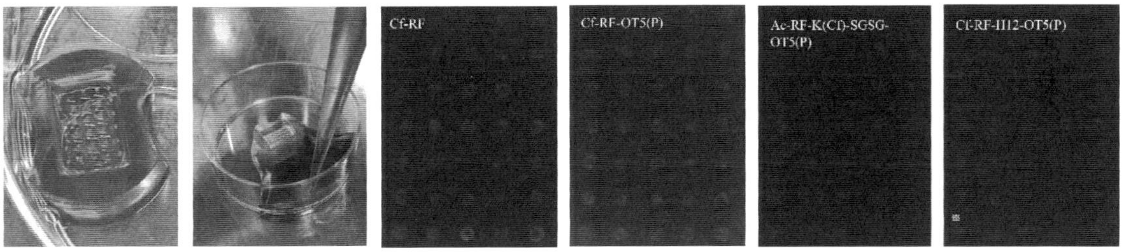

Figure 3. Penetration studies on agarose dish-based A431 spheroids, treatment applying 12.5 μM Cf-peptides, 3h.

The tandem peptides' penetration (with low UC_{50} values) was evaluated on spheroids as simple *in vitro* tissue-mimicking milieu, and all tandems demonstrated fair penetration ability (Figure 3).
These peptides are promising carrier candidates with variable intracellular trafficking depending on their sequence.

Acknowledgements

Grants OTKA 142904; VEKOP-2.3.3-15-2017-00020; EFOP-1.8.0-VEKOP-17-2017-00001; 2018-1.2.1-NKP-2018-00005; TKP2020-IKA-05; Gedeon Richter Talentum Foundation (LL).

References

1. Bősze. S., et al. *Biomolecules*, **10(5),** 721. (2020), https://doi.org/10.3390/biom10050721
2. Horváth, L.B., et al. *Eur J Pharm Biopharm.*, **174**, 111-130. (2022), https://doi.org/10.1016/j.ejpb.2022.03.009
3. Baranyai, Z., et al. *J Med Chem.*, **64(6),** 2982-3005. (2021), https://doi.org/10.1021/acs.jmedchem.0c01399
4. Horváth, L.B., et al. *ACS Omega.*, **9(15),** 16927-16948 (2024), https://doi.org/10.1021/acsomega.3c05727
5. Lamiable, A., et al. *Nucleic Acids Research*, **44**, W449-454 (2016), https://doi.org/10.1093/nar/gkw329

Proceedings of the 37th European Peptide Symposium
Michal Lebl, Editor
European Peptide Society, 2024
http://doi.org/10.17952/37EPS.2024.P2247

Synthesis and application of β-amino γ-lactam (Bgl) residues

Charity D. Yongo-Luwawa[1] and William D. Lubell.[1]
[1]Department of Chemistry, Université de Montréal. Montréal, H2V 0B3, Canada

Introduction

α-Amino γ-lactam (Agl), so-called Freidinger-Veber lactam residues have garnered significant interest in the realm of peptide mimicry due to ability to introduce backbone constraint (Figure 1)[1,2]. In peptides, the Agl residue prefers the *i*+1 position of β-turn conformers and can thereby stabilize a bioactive conformation that is commonly found in natural peptides[1,2]. Compared to Agl residues, β-amino γ-lactam (Bgl) analogs have been less frequently explored in spite potential to favor turn geometry. For example, Bgl residues have been utilized in the study of growth hormone-releasing peptide-6 (GHRP-6, H-His-D-Trp-Ala-Trp-D-Phe-Lys-NH_2) to gain insight into a turn conformer about the Ala[3] residue[3]. Moreover, a Bgl scan of 101.10 (H-D-Arg-D-Tyr-D-Thr-D-Val-D-Glu-D-Leu-D-Ala-NH_2) illustrated the relevance of turn conformers for interleukin-1 receptor modulation[4]. Highlighting utility for stabilizing biologically active peptide conformers, such studies spur interest in synthetic methods for making Bgl residues. Previously, Bgl peptides have been synthesized using aspartate-derived cyclic sulfamidates as bis-electrophiles[4]. Towards an alternative method for Bgl peptide synthesis, commercially available (*S*)-(–)-β-hydroxy-γ-butyrolactone has now been employed as precursor.

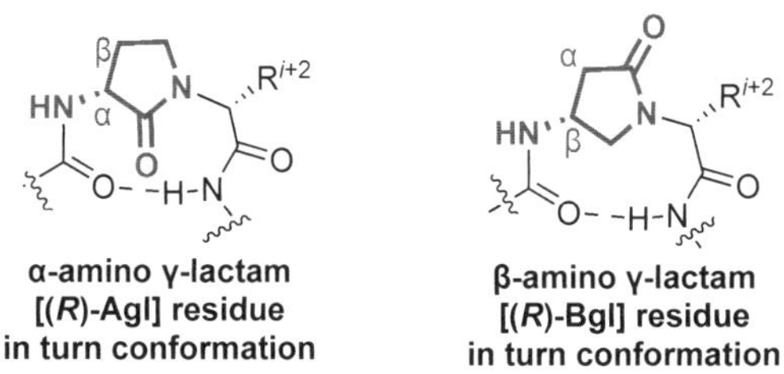

Figure 1. Agl and Bgl residues at i+1 position of β-turn conformations

Results and Discussion

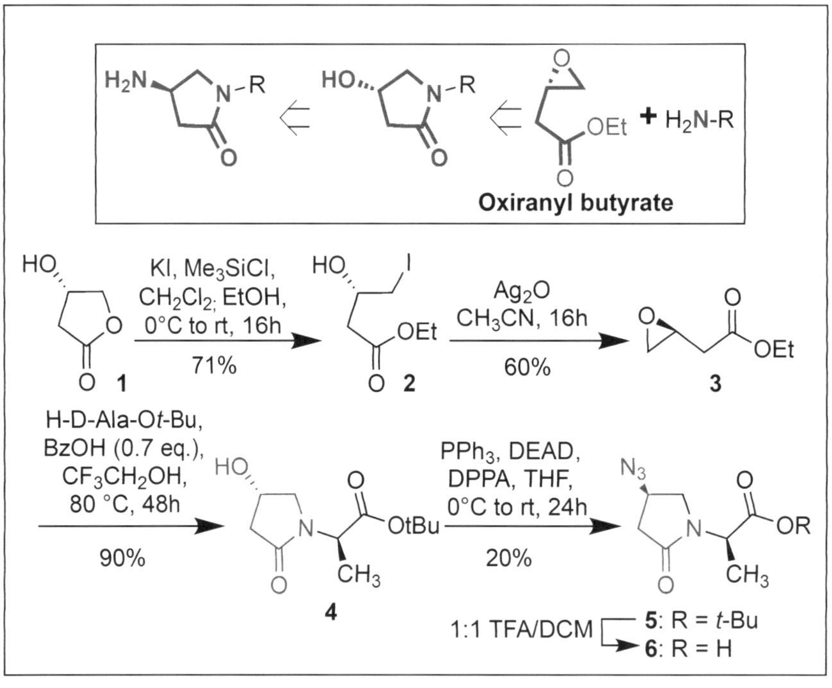

Figure 2. Conceptual and actual applications of oxiranyl butyrate in Bgl residue synthesis.

 http://doi.org/10.17952/37EPS.2024.P2247

(*S*)-(–)-β-Hydroxy-γ-butyrolactone undergoes a ring-opening reaction in the presence of potassium iodide, trimethylsilyl chloride and ethanol to provide iodo ester **2** in 71% yield[5] (Figure 2). Oxiranyl butyrate **3** is prepared from iodo alcohol **2** on treatment with silver oxide in acetonitrile in 60% yield[5]. In proof-of-concept studies, oxiranyl butyrate **3** was employed as a bis-electrophile in the alkylation and acylation of D-alanine *tert*-butyl ester. Employing benzoic acid as catalyst in trifluoroethanol at 80°C for 48 h gave β-hydroxy lactam **4** in 90% yield[6]. An initial attempt to displace the alcohol by azide using diethyl azo dicarboxylate (DEAD), diphenyl phosphoryl azide (DPPA) and triphenyl phosphine in THF provided Bgl analog **5** (20% unoptimized), which after removal of the *tert*-butyl ester with trifluoroacetic acid in dichloromethane is suitable for incorporation into peptides. The scope of this approach is currently being developed with interest in expanding the synthesis of Bgl analogs in peptides including GHRP-6 and 101.10 to study further biological activities.

Acknowledgments

We thank the Canadian Institutes of Health Research (CIHR) for project PJT-190158, the Natural Sciences and Engineering Research Council of Canada (NSERC) for funding Discovery Research Project RGPIN-2019-04079, and the Fonds de Recherche du Québec - Nature et Technologie (FRQNT) for support of the Centre in Green Chemistry and Catalysis (CGCC, FRQNT-2020-RS4-265155-CCVC). The authors acknowledge the assistance of members of the Université de Montréal facilities: Dr. P. Aguiar (NMR spectroscopy) and Dr. A. Fürtös (mass spectrometry). CYL is grateful for a doctoral fellowship from the FRQNT.

References

1. Freidinger, R. M.; Veber, D. F.; Perlow, D. S.; Brooks, J. R.; Saperstein, R., Bioactive conformation of luteinizing hormone-releasing hormone: evidence from a conformationally constrained analog. *Science* **1980,** *210* (4470), 656-665.
2. St-Cyr, D. J.; García-Ramos, Y.; Doan, N. D.; Lubell, W. D., Aminolactam, N-Aminoimidazolone, and N-Aminoimdazolidinone Peptide Mimics. In Peptidomimetics I; Springer, 2017; pp 125–175.
3. Boutard, N.; Jamieson, A. G.; Ong, H.; Lubell, W. D., Structure–Activity Analysis of the Growth Hormone Secretagogue GHRP-6 by α- and β-Amino γ-Lactam Positional Scanning. *Chem. Biol. Drug Des.* **2010,** *75* (1), 40-50.
4. Jamieson, A. G.; Boutard, N.; Beauregard, K.; Bodas, M. S.; Ong, H.; Quiniou, C.; Chemtob, S.; Lubell, W. D., Positional Scanning for Peptide Secondary Structure by Systematic Solid-Phase Synthesis of Amino Lactam Peptides. *J. Am. Chem. Soc.* **2009,** *131* (22), 7917-7927.
5. Sunnapu, R.; Rajendar, G., A Concise Stereoselective Total Synthesis of Methoxyl Citreochlorols and Their Structural Revisions. *Eur. J. Org. Chem.* **2021,** *2021* (11), 1637-1642.
6. St-Cyr, D. J., Jamieson, A. G.; Lubell, W. D., α-Amino-β-hydroxy-γ-lactam for constraining peptide Ser and Thr residue conformation. *Org. Lett.* **2010**, *12* (8), 1652–1655.

Proceedings of the 37th European Peptide Symposium
Michal Lebl, Editor
European Peptide Society, 2024 *http://doi.org/10.17952/37EPS.2024.P2294*

A comprehensive structural study on AT1 Receptor Variants: Towards the Design of Potent Biased Agonists

Malihe Hassanzadeh(1), Audrey Collette(1), Brian Holleran(1), Stéphane Laporte(2), Richard Leduc(1) and Pierre-Luc Boudreault(1)*

(1) Department of Pharmacology-Physiology, Faculty of Medicine and Health Sciences, Institute de Pharmacologie de Sherbrooke, Université de Sherbrooke, 3001 12th Ave North, Sherbrooke, Qc, J1H 5N4, Canada
(2) Department of Pharmacology and Therapeutics, McGill University, Montréal, Québec, H3G 1Y6, Canada

The angiotensin II type 1 receptor (AT_1R) is a crucial component of the renin-angiotensin system (RAS), which plays a significant role in regulating blood pressure, and cardiovascular health [1]. Biased agonism is an emerging concept that activates specific signaling pathways of G protein-coupled receptors (GPCRs). It involves selectively targeting the receptor with a ligand that stimulates one pathway while inhibiting others, leading to more precise treatments with fewer side effects. To gain a deeper understanding of GPCR functional selectivity at a molecular level, this study focuses on the use of AT_1R mutants and biased agonists. The activation of the AT_1R by different agonists leads to different signaling pathways which is known as functional selectivity [2]. Understanding this mechanism can help to develop more effective and targeted treatments for hypertension and other cardiovascular diseases. To assess the activation of Gq and the recruitment of β-arrestin in response to the stimulation of AT_1R with angiotensin II (AngII), [Sar^1Ile8], AngII [Sar^1LBpa^8]AngII, and [Sar^1DBpa^8]AngII ligands, we opted to substitute position W253$^{6.48}$, also referred to as the rotamer toggle switch, with alanine (W253A-AT_1R). The activation of G_q and recruitment of β-arrestin were quantified using BRET-based biosensors. Our results indicate that the mutant W253A-AT_1R was able to activate Gq and recruit β-arrestin when stimulated with AngII. However, upon stimulation with [Sar^1Ile^8]AngII or [Sar^1DBpa^8]AngII, which are β-arrestin-biased ligands on AT_1R, W253A-AT_1R mutant failed to recruit β-arrestin. Interestingly, [Sar^1LBpa^8]AngII could activate β-arrestin in the W253A-AT_1R mutant background.

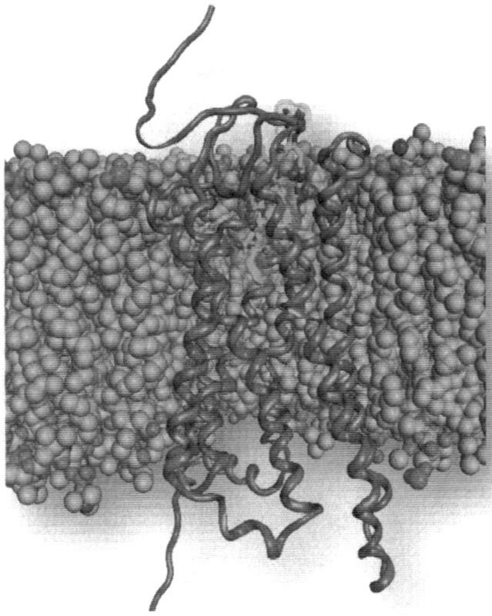

This study proposes significant interactions exist between the mutant W253A-AT_1R and specific determinants of the different peptide ligands, which are involved in stabilizing the receptor in a conformation capable of recognizing β-arrestin. To investigate the structural dynamics of the native AT_1R and W253A-AT_1R mutant receptor, we utilized molecular dynamics (MD) simulations. Our study incorporated interactions with the native endogenous ligand and the three modified peptides. Our simulations revealed distinct conformational landscapes for each receptor-ligand pairing, with particular emphasis on regions of flexibility and interaction. Significantly, these unique conformational tendencies, lead to a better understanding of the molecular determinants responsible for selecting a given biased pathway. This information provides opportunities to design biased agonists with great therapeutic promise, as they may offer the potential for enhanced efficacy and reduced side effects.

References:

1. Van Haaster, McDonough, Gurley, Curr Opin Nephrol Hypertens, 2018, 27, 1-7.
2. Wingler, Skiba, McMahon, Staus, Kleinhenz, Suomivuori, Latorraca, Dror, Lefkowitz, Kruse, 2020,367, 888-892.